Praise for SQUEEZED

One of CSPAN's Most Important Books of the Year

One of *Time*'s Best New Books to Read This Summer

One of *Nylon*'s Great Books to Read This Summer

Named a best book of the month by the *Christian Science Monitor*, *Library Journal*, and *Bitch* magazine

"An eye-opening look at the forces that make it harder than ever for the middle class to survive." —*People*

"Brilliant—a keen, elegantly written, and scorching account of the American family today."
—Barbara Ehrenreich, author of *Nickel and Dimed*

"In a nation beset by income inequality and riven by social and cultural conflict, the traditional conception of the quiet contentment of middle-class American life appears to be on the wane. . . . Alissa Quart . . . lucidly demonstrates that for many, the dream of such satisfaction is increasingly out of reach." —*Boston Globe*

"It's not often that you can call a densely reported work of literary non-fiction about economic inequality a riveting page-turner, but *Squeezed* is just that." —*Guardian*

"Eloquent and passionate. . . . In deeply etched portraits of struggling professionals, Quart evokes how soaring costs and hostile social policy have trapped middle-class families in quicksands of debt and emotional stress." —*Forward*

"The issue is overwhelmingly structural and social, not individual or moral. We haven't failed; Capitalism has failed us. As Quart reminds her reader—and as every story in the book is meant to illustrate—the economic bind we find ourselves in cannot be solved by personal discipline or better financial decisions."

—*New York Times Book Review* (cover)

"Think of Alissa Quart's new book . . . as 'What to Expect When You're Expecting Under Late Capitalism.' Of the more than 50,000 books listed on Amazon under 'Parenting,' few engage as deeply with the economic pressures today's parents must navigate."

—*In These Times*

"*Squeezed* captures well the toxic combination of American individualism and the disrupted evolution of particular professions that has left millions of millennials in a more fragile financial condition than they expected would be their lot in life."

—*Washington Post*

"If Ehrenreich's *Nickel and Dimed* changed the conversation about poverty-wage jobs nearly 20 years ago, *Squeezed* . . . explains the financial insecurities not only of the poor, but also of a middle class precariously balanced on the edge."

—*The American Prospect*

"*Squeezed* examines the deteriorating fortunes of the middle-class[, illustrating] how life in a once-secure stratum has come to resemble the endlessly anxious existence of those in the rungs below."

—Ginia Bellafante, *New York Times*

"[Quart] shares familiar stories of economic frustration as well as hard evidence for the causes of it. It's an often tough but deeply empathetic call to action, one that exists in the real world of family, work, debt and even dreams."

—Salon

"*Squeezed* captures the dazed uncertainty of a post-recession generation of would-be parents for whom stagnant wages and ever-rising housing costs make them can't-be ones. . . . Quart [has] a knack for immersive, in-depth reporting, as well as an often-bruised sense of unlikely optimism."

—*Bitch* magazine

"Quart is a sympathetic listener, getting people to reveal not just the tenuousness of their economic situations but also the turbulence of their emotional lives. . . . We could all use her expert guidance through the maze." —*New York Times*

"Lucidly recounts . . . wrenching stories of economic hardship, while meticulously deconstructing some of the prevailing myths about middle-class life in the United States. . . . *Squeezed* stands out for its insightful analysis of class dynamics in the United States." —*Star Tribune* (Minneapolis)

"*Squeezed* is at its absolute best when Quart combines her excellent reporting, rich with the perspective of that 'Middle Precariat,' with a willingness to more openly and sharply decry the structural conditions that are giving way to such misery." —PopMatters

"The problem, Quart argues, isn't that Americans are managing their money badly or are lazy, but that the economy has fundamentally changed." —CBS News

"Quart captures how middle-class American families are struggling to attain the standard of living once enjoyed by their parents. . . . [She] argues that higher earners, like most Americans, contend with income disparity and the extreme wealth enveloping metro regions." —*The American Conservative*

"The stories of a falling down middle class reflect a felt experience of anxiety that is often lost in data-driven tales of recession and recovery." —*Financial Times*

"A powerful book. It enlightens and provides context in real-life stories of people trying to make it in America, not to strike it rich but to live in relative freedom from financial brinkmanship. . . . The color of your collar, or your red or blue state, doesn't matter. Quart reveals in this work that we're in this together."

—*Martha's Vineyard Times*

"There are books you read for pleasure. Then there are books you need to read to be well-informed. *Squeezed* falls into the latter category. . . . Quart offers some solutions, although what makes the book compelling are the stories. You meet people, who for various reasons—health issues, child-care needs, crushing student loans, underemployment or job displacements—are just barely hanging on to middle-income status. They are a job or health crisis away from slipping even further."

—*Washington Post* Color of Money Book Club

"A devastating report on middle-class American families struggling to stay afloat. . . . Quart eloquently relates these families' psychological and socioeconomic predicaments, and amplifies those personal stories with research confirming significant and growing income inequality."

—The National Book Review, "5 Hot Books"

"A compassionate, thoughtful examination of the economic struggles faced today by America's middle class."

—*Christian Science Monitor*, "10 Best Books of June"

"Reminiscent of Barbara Ehrenreich's *Nickel and Dimed* . . . will resonate with those feeling squeezed, and inform those who are not." —*Library Journal*

"A thoughtful, enlightening and painful analysis of the ever-growing divide in the American economy." —*BookPage*

"Puts plain the economic predicament of the middle class in eloquent and heart-wrenching vividness." —Electric Literature

"It's okay to feel angry after reading this. In fact, you should. Let this inspire you to protest the rampant inequality in this country, which is dooming millions of people to lives of desperation and destitution and despair."

—*Nylon*, "46 Great Books to Read This Summer"

"Well-written, wide-ranging, and vital to understanding American life today." —*Kirkus Reviews* (starred review)

"*Squeezed*, like *Nickel and Dimed* . . . is worth reading if you're invested in better understanding poverty in America."
—*Bitch* magazine, "15 Books Feminists Should Read in June"

"Examines the lives of many American families whose middle-class dreams are increasingly out of reach." —*Missourian*

"[Quart's] ambitious, top-tier reportage tells a powerful story of America today." —*Publishers Weekly*

"We can only blame people for so long for not living within their means. Eventually, we have to study the means."
—*Globe and Mail*

"A disciplined journalist, Quart [frames] her facts and figures with unsettling personal stories of financial ruin. . . . What stays with you is the precariousness and stress to which today's capitalism subjects millions of even relatively privileged people." —*Plough*

"The stories Quart passionately reports are heart-wrenching. They are also a clear warning that this nation is heading in a perilous direction. *Squeezed* is journalism at its best: exploratory, visceral, and searching for answers. An important work to which attention should—and must—be paid."
—David Corn, author of *Russian Roulette: The Inside Story of Putin's War on America and the Election of Donald Trump*

"Quart's investigation, written with the elegance of a literary novel, forces us to examine the grave consequences of an economic structure that has crushed the very people it claims are at the heart of the American dream."
—Jeremy Scahill, author of *Blackwater* and *Dirty Wars*

"What Alissa Quart does so beautifully is weave together textured, compelling portraits of individual families with big ideas. Read this important book to understand the challenges your own family faces in parenting, housing, planning for the future—and read it to find out what to do about them!"

—Peggy Orenstein, author of *Girls & Sex* and *Don't Call Me Princess*

"Alissa Quart is a modern James Agee. *Squeezed* gets deep inside the increasingly perilous financial lives of American families, showing that they are collateral damage of our disappearing government. A damning, necessary, and intensely vital book."

—Helaine Olen, author of *Pound Foolish: Exposing the Dark Side of the Personal Finance Industry*

"If you've ever felt the pinch of financial anxiety—and chances are you have—read this book now. Alissa Quart will help you realize that you're not alone and it's not your fault. *Squeezed* is profound; a sweeping, blistering portrait of hard-working people from all walks of life. It's a rousing wake-up call that also points the way forward to a more equitable, expansive future."

—Astra Taylor, author of *The People's Platform: Taking Back Power and Culture in the Digital Age*

"A vivid prose stylist, Quart makes powerfully real what happens when those who were once middle-class can now only window shop for the American Dream."

—Arlie Hochschild, author of *Strangers in Their Own Land: Anger and Mourning on the American Right*

SQUEEZED

also by ALISSA QUART

BRANDED

HOTHOUSE KIDS

REPUBLIC OF OUTSIDERS

MONETIZED
(POETRY)

SQUEEZED

WHY OUR FAMILIES CAN'T AFFORD AMERICA

ALISSA QUART

An Imprint of HarperCollins*Publishers*

Some of the reporting and thinking in this book appeared in a substantially different form in magazines and news sites, namely the *New York Times*, *The Nation*, *Elle*, the *Guardian*, *The Atlantic*, and *Pacific Standard*. I thank the great editors at these publications that I worked with over the last five years as I developed the ideas and characterizations that compose these pages. The names of a few individuals in one chapter of this book have been changed to protect their privacy.

A hardcover edition of this book was published in 2018 by Ecco, an imprint of HarperCollins Publishers.

FIRST ECCO PAPERBACK EDITION PUBLISHED 2019.

Designed by Renata De Oliveira

The Library of Congress has catalogued a previous edition as follows:

Names: Quart, Alissa, author.
Title: Squeezed : why our families can't afford America / Alissa Quart.
Description: New York : Ecco, 2018.
Identifiers: LCCN 2017056972 (print) | LCCN 2018013105 (ebook) | ISBN 9780062412270 (ebook) | ISBN 9780062412256 (hardback) | ISBN 9780062412263 | ISBN 9780062847904 | ISBN 9780062847911
Subjects: LCSH: Middle class—United States—Economic conditions. | BISAC: SOCIAL SCIENCE / Sociology / Marriage & Family. | POLITICAL SCIENCE / Public Policy / Economic Policy. | POLITICAL SCIENCE / Public Policy / Social Policy.
Classification: LCC HT690.U6 (ebook) | LCC HT690.U6 Q37 2018 (print) | DDC 305.5/50973—dc23
LC record available at https://lccn.loc.gov/2017056972

ISBN 978-0-06-241226-3 (pbk.)

19 20 21 22 23 LSC 10 9 8 7 6 5 4 3 2 1

To my daughter, Cleo

CONTENTS

INTRODUCTION

Michelle Belmont's debt haunted her. It was almost unspeakable, but it was a raw relief when anyone asked her about it. She wanted people to hear about her life as she lived it, how her debt trailed her like a child's monster, how it was there when she went to the supermarket, to her son's day care, and home to her one-bedroom apartment.

It began as it often does, with the student loans for the college her parents back home in Georgia thought would ensure the right future. Then there was the money she borrowed for her master's of library science degree. A bit later, when baby Eamon came along, she and her husband owed over $20,000 in hospital bills as well. What was shocking were the price tags, just for normal things, like Michelle's labor and her overnight stay. She had required a few days extra at the hospital: Eamon had been born weighing ten pounds, thirteen ounces, and she had pushed that hefty creature for five hours.

"I thought that insurance helps you get by," Michelle told me. "But my husband had a really cheap insurance, and you get what you pay for."

Then the debt shadow monster just grew. Eamon developed a fever of 103 degrees and had to go back to the hospital.

There were two years of surgeries. The bills piled up on the kitchen table. Michelle tried to pay them off, for fear of getting refused treatment later, but then she stopped opening the envelopes. They were different colors. They demanded payment now or legal action, in screaming capital letters. She saw herself on trial, in court, explaining why she had nothing in her account. Her debt was six figures and growing.

The couple had struggled before they had their baby, Michelle said, but then "it got astronomically insane after Eamon was born. We always had money for food before, but now it's, 'How are we going to eat?' I'll borrow from one credit card bill to pay that other credit card bill. I can't find rent money each paycheck, and we make a decent salary between us."

Michelle Belmont was fighting to stay middle-class. She hoped to train herself into a career of certitude—to become a technological librarian, to set up her future. But the costs were beyond what she ever imagined, and she grew more vulnerable. Meanwhile, the squeeze tightened. The Belmonts lived in a modest one-bedroom apartment in Minneapolis that she and her husband paid $1,300 per month to rent. Minneapolis, with its supposed hipster status and so-called Midwest Modern food and furniture and textiles, was only getting more expensive for Michelle. When I first spoke to her, it seemed unlikely that the Belmonts would ever be free of debt.

"That requires nothing bad to happen," Michelle said, almost laughing.

But bad things do happen.

When I first spoke to Michelle, her concerns were not abstract to me. Back then, I had recently given birth to my daughter. And it wasn't until I had my own child that I quickly realized that I too had entered the falling middle-class vortex. My girl was born face-first—sunny-side up, as they say—her unblinking stare promising new joy and terror. Her cries soon became the soundtrack of the anti-romantic comedy of our lives. My husband and I wound up with an unexpected $1,500 bill after her birth that we hustled to pay; most Americans owe even more, an average of around $5,000. Although we managed to avoid the financial perils that many of the people you will meet in this book experienced—partly because of the wonder of having a New York City rent-stabilized apartment—we did go through a few years of fiscal vertigo. We had been freelance writers for most of our careers, but by the time my daughter arrived this was no longer a stable line of work for the majority of its practitioners, including us. And now we had day-care costs and hospital bills. We started to search for jobs with regular pay, regular hours, and health insurance.

My husband was already fifty, and it turned out that our years of relative liberty—of "doing what we loved"—had finally exacted a price. When our daughter was four months old, it got even worse. We first hired a nearly full-time sitter and most of my own take-home earnings as an editor went directly to her. Eventually, my earnings also flowed to my daughter's cheerfully boho day care (even though, paradoxically, all the caregivers were most likely themselves just scraping by, despite their loving and primary-color-bright attentions). Again, given the larger field of suffering, our family's worries were relatively low-key. But still we yearned for more of a social mesh to keep us afloat. At the time, we felt like startled nocturnal animals. Subsidized day care had done so much for us. And how much would it have done for people who did not have as many choices as we did?

Eventually, my husband found a full-time editorial job, and so did I. Perhaps not so coincidentally, mine was as director and editor of a journalism nonprofit devoted to supporting reporting on inequality by a good number of reporters who had themselves fallen on truly hard times. I continue to spend my days editing these narratives.

Through these full-time positions, our family was saved from tumbling out of our class position—at least for now. But even after we found ourselves in momentary safety, I couldn't shake the self-blame. Despite our encroaching middle age, *we had not planned ahead*, I thought. I felt juvenile, but also suspected that the game was rigged—that unlike me, the very wealthy who now filled the city of my birth and worked in finance didn't lacerate themselves for small missteps.

This personal experience was partly how I arrived at what was to become the mantra of this book: *It's not your fault*. It seems key to me—to recognize that feeling in the red or on the edge *isn't all your personal problem*. And while some psychological analysis or boosts may help, the problem of not being able to afford to live in America can't be cured by self-help mantras. It can't be mended simply by creating a résumé that utilizes several colors of printer ink or a regimen of cleansing green juices. The problem is systemic.

Squeezed is the story of this psychological and socioeconomic predicament. Being squeezed involves one's finances, one's social status, and one's self-image. The middle class I refer to in these pages is a group defined by more than just money: it also leans on credentials, education, aspirations, assets, and, of course, household income. In the United States, the middle class is the group of working people who, according to a May 2016 Pew survey, with a yearly household income for a family of three ranging from $42,000 to $125,000 in 2014, make up 51 percent of U.S. households. Michelle Belmont and her family were in the middle class, and they were squeezed.

The middle-class families running furiously and breathlessly just to find themselves staying in place are a large and varied coterie. It includes highly educated workers like lawyers, professors, teachers, and pharmacists, professionals who never expected to be in this situation—often feeling cast aside by a system that seems stacked against them. Their prospects for the future, given the rise of robots and automation within their professions, which you will read about later in this book, are likely to dim even further.

According to a *Washington Post*/Miller Center poll, 65 percent of all Americans worry about paying their bills—as the parents I've interviewed, murmuring anxiously at their dining room tables, can attest. One reason for this anxiety is that middle-class life is now 30 percent more expensive than it was twenty years ago; in fact, in some cases the cost of daily life over the last twenty years has doubled. And the price of a four-year degree at a public college—one traditional ticket to the bourgeoisie—is nearly twice as much as it was in 1996. The cost of health care has almost doubled in that twenty-year period as well. And rent, not to mention homeownership, has also become substantially more expensive, though not quite to the same horrifying level as education and medical care. Meanwhile, the ongoing decimation of unions and employees' rights continues, with pensions and minimal benefits fading. Unstable working hours are increasingly common too, making child care, always a high personal expense for families, all the harder to arrange and even more expensive while further testing family cohesion. And the squeeze on the middle class has an element of gender bias as well. It's no accident that many of the people you'll meet in this book are female. Although there are other reasons why so many of the characters of *Squeezed* are from the distaff side, there is one quite simple reason: motherhood is a disadvantage in the work world, with mothers statistically earning less than their male or childless peers. But fathers are harmed too: if they strive to more evenly balance their careers and their families, they may be stereotyped at work as "weak." And if they go into traditionally female caring professions—where most of the employment growth is these days—they may receive the "traditional" female lower pay.

I call this just-making-it group "the Middle Precariat," after the *precariat,* a term first popularized six years ago by the economist Guy Standing to describe an expanding working class burdened with temporary, low-paid, and part-time jobs. My term, the Middle Precariat, describes those at the upper end of that group in terms of income. Its membership is expanding higher and higher into what was traditionally known as the solid bourgeoisie. These people believed that their training or background would ensure that they would be properly, comfortably middle-class, but it has not worked out that way. Their labor has also become inconstant or contingent—they do short-term contract or shift work, as well as unpaid overtime. They also do unpaid shadow work, like adjunct professors putting together packets for their classes off the clock, in contrast to their tenured colleagues. And it's worse for the Middle Precariat of color, which typically has much less retirement security and ability to pay college tuition.

Like the classic precariat, the Middle Precariat has lost the narrative of their lives and futures. Who are they and what will they become? Their income has flatlined. Many are "fronting" as bourgeois while standing on a pile of debt. There are many culprits for the straits in which they find themselves—most crucially growing income inequality, or as the business TV shows like to call it euphemistically, as if to deny their role in creating it, "disparity." The United

States is the richest and also the most unequal country in the world. It has the largest wealth inequality gap of the two hundred countries in the *Global Wealth Report* of 2015. And when the top 1 percent has so much—so much more than even the top 5 or 10 percent—the middle class is financially and also mentally outclassed at each step.

Behind the proverbial velvet curtain—or midrange eggshell-colored Roman blinds—these parents are desperately holding on to their status and trying to keep up appearances.

This is a true historical shift. When I posted on Facebook that I couldn't afford the relatively modest life my academic parents had, many friends added their own stories about how their income goes to rent and child care—the latter often siphoning off up to 30 percent of a family's earnings. The cost of having kids can seem like Eric Carle's *The Very Hungry Caterpillar:* just like the caterpillar in the classic children's book, your child eats up every dollar you earn. The cost of child care reflects a reality principle: according to a 2016 study by the Equality of Opportunity Project, Americans born in the 1940s had a 92 percent chance of making more money than their parents did at age thirty. Those born in the 1980s have around a 50 percent chance of earning more than their parents. (In the Midwest, as the *New York Times* reported, the odds are less than half.)

When I was a young child, professional aspiration was synonymous to me with the clatter of my mother's high-heeled boots as she went off to teach each 1970s weekday morning, carrying her graded blue books under her arm. Each day was concluded when my exhausted mother picked me up late at the very end of after-school and took me home for a dinner of spaghetti and meatballs. Yet despite the evident effort they put in, my parents, college professors, had health insurance and the promise of pensions and Social Security. In their younger days, there were ample employment opportunities and cheaper rents in metropolitan areas. They could afford some extras that would strain a similar family today: out of their wages from teaching at a college, I received ice skating lessons. I was sent to a New York City private school, and we went on long vacations at the shore, where I could buy a kite in the shape of a butterfly and maybe collect wild plums on the dunes. They weren't alone. "Middle-class" used to mean having two children and sending them to high-quality public schools, or even occasionally to private schools. It meant new brown Stride Rite Mary Janes with little purple and silver flowers when the old shoes were pinching the toes. It meant homeownership—not for us, but for others like us. Nothing fancy, but a proper ranch house with a garage. It meant weekends off with your family, sometimes spent at a matinee at a "movie palace," or a play thanks to a theater subscription, and workdays that ended at six so that the family all ate dinner together. And of course, it meant saving money as well as being able to pay for the children's college education.

For the American middle class now, these markers of middle-class life are less and less common. The middle class is endangered on all sides, and the promised rewards of belonging to it have all but evaporated. This decline has also led to a degradation of self-image. Before the 2008 crash, only one-quarter of Americans viewed themselves as lower-class or lower-middle-class. Even those who were struggling tended to view their problems as temporary. No longer. After the recession of 2008—which, though caused by the financial crash, could actually be said to have exposed or congealed decades of social class separation and downward mobility, since the Reagan era—a full 40 percent of Americans viewed themselves as being at the bottom of the pyramid. For the first time since pollsters had asked this question, fewer than half of those interviewed said that they were middle-class—only 44 percent, according to a Pew study. Meanwhile, the wealthy—with "wealth" here defined as assets minus debt—stand in stark relief to the Middle Precariat. A 2014 Russell Sage Foundation report puts the net worth of the top 5 percent at $1.3 million. The incomes of the top 1 to 5 percent have grown explosively in the past three decades, while the incomes of so many others have stagnated.

For the median family of color, that wage and wealth stagnation can be pretty dire. In a study published in 2017 by the organizations the Institute for Policy Studies and Prosperity Now (full disclosure; IPS is the fiscal sponsor of my organization, the Economic Hardship Reporting Project), the median wealth—assets minus debt—of white households is now over sixty-eight times higher than that of black households. For black families, the median was just $1,700.

The 2017 tax bill will likely only make these numbers even worse for many Americans. But this so-called tax reform is only the most recent example of how income inequality is written into the law of the land.

If you are an American working parent dealing with all of these stresses, you may feel like you are betting against the house and the house is always winning. Yet most of the parents I spoke to blamed only themselves, not a system stacked against them.

In *Squeezed,* you will meet a professor on food stamps in Chicago, an unemployed restaurant manager in Boston, and a nanny in New York City betrayed by the American Dream, and you will even hear about pharmacists who lost their jobs to a robot in Pittsburgh. They are people on the brink who did everything "right," and yet the math of their family lives is simply not adding up. Some are just getting by. For others, something happened and they tumbled down and never got back up.

For mothers in particular, this situation can be something I call the "class ceiling," the intersection of the "glass ceiling" that stymies workingwomen's careers and the result of the myriad injuries of social class.

This book hopefully illuminates the lives of the struggling middle class and offers strategies that may help. As these families struggle to preserve, or even simply to attain, a middle-class life, they do so in spite of, not because of, today's America. Here are their stories.

1

INCONCEIVABLE

PREGNANT AND SQUEEZED

It should have been her heyday. Daniela Nanau was in her thirties and had been working at the law firm for ten months as an associate. She was excited that her new partners were so celebrated in her field, employment law, and she adored going to the tasteful office in New York City, lined with art. She worked closely with her boss and believed that they liked each other. She even felt like they were capable of a rare sort of "professional mind-meld," as she put it, and that they had the same kind of personality.

Then Nanau became too queasy to get through her morning commute: some days, she'd sit down on a bench in the small green space near her bus stop in Queens, New York, and lose her place in the bus queue. Like a suburban Sisyphus, by the time she'd gather the strength to stand up, the crowd would have outpaced her and she'd have to go to the back of the long line. When she finally got to the exposed brick room of her office space, she had to steady herself and could barely sit up in her desk chair. She felt so weak that she went to see an oncologist. Did she have cancer? When the results from her blood test came back, he asked, "Don't you know what's going on in your life?" She was pregnant.

If this were a Hallmark card, or an era when mothers were truly the "angels" in their houses, Nanau would probably have been elated. She would have had nothing to do but decide whether her fetus was now the size of a lychee or a grapefruit. Instead, she was in physical and emotional agony. She was, after all, her family's main earner, the one with the graduate degree. Like her, Nanau's husband had started out working in politics in Washington, D.C., but now he remodeled houses, plying his trade with power tools and putty knives, which all took up serious space in their home. His work was not very profitable. If Nanau got laid off or didn't get another job, her family wouldn't be able to make their mortgage payments.

As she worried into the night, sitting on a mustard-yellow mohair couch, a worn 1950s heirloom from her German grandparents, she remembered what partners at other law firms had told her in passing at luncheons. If she wanted to survive in the field, she couldn't have children until her forties. She tried to push all of these intrusive thoughts out of her mind and decided to tell her boss she was pregnant. Immediately afterward, she said, he ignored her and refused to talk to her for a week.

She was pregnant yet shedding pounds from her already slim frame. She soon realized that her boss had turned on her irredeemably because of her illness and her lateness to work, but also, she thought, he seemed to believe that her pregnancy would make her even weaker. She believed that her boss was shunning her for wanting to have children in the first place; ignoring her, he spoke to her harshly when he spoke to her at all. She was afraid to complain or sue, however, as her own clients did, because the legal community in their field in New

York City was small. Her husband knew that during the ten months she'd put in at the firm she routinely worked in the office until midnight once a week, sometimes without dinner. He told her to leave the job immediately. Nanau began to look for other work.

When Nanau eventually quit, she didn't tell her employer why. She was one of the lucky ones, as it turned out. Nanau was able to climb the "maternal wall" and escape the lingering impact of bias. After she left the job with the "bad boss," she was able to get another legal job that was even better paid, she said. And that made all the difference for her personally. Over the next few years, her employment discrimination practice was shaped by the prejudice she had experienced herself, but she kept her personal resentment buried.

Nanau and the women whose rights she defends are not alone. The way we treat expectant women is a symptom of how little American businesses and legislators care about *care*. Another symptom: pregnancy discrimination cases are on a massive upswing. In 2016, a report published by the Center for WorkLife Law found that so-called family responsibilities discrimination cases had risen 269 percent over the last decade, even though the number of federal employment discrimination cases as a whole had decreased. And women who said they weren't hired because of their pregnancies were responsible for 10 percent of all discrimination claims to the Equal Employment Opportunity Commission (EEOC) in 2011, a substantial rise.

The jump in family discrimination cases reflects in part the rise in the numbers of employees, both male and female, who now have caregiving responsibilities. The number of parent employees who are part-time caregivers for their children has increased, partially due to an expanded female workforce. Workplaces have not changed enough. In this country, pregnancy—and as we shall see later, being a parent at all—can be a professional hazard. Nanau recounted that the female workers for whom she prosecuted cases were not given chairs to sit in while they worked behind cash registers into their sixth month of pregnancy. The white-collar workers she represented were harassed in subtler ways: They were given more work than they could handle, offered "friendly" suggestions by their colleagues about work-life balance, or simply needled about everything from their clothes being too tight to whether they suffered from postpartum depression. Her voice rose and her face flushed. So many others had it far worse than she and her friends, she said. How could babies be the culprits here? How could we live in a society that scorns pregnancy? When you break it down, society's attitude toward pregnancy and children is bizarre, surreal, and much like the disregard for so much of nature itself.

This book is about families who are pressed economically. The vise in which they are trapped wouldn't be as powerful in the first place, however, if we properly valued care. Pregnant women and children, two populations romanticized in a retrograde way by greeting cards and Republican candidates, receive kind glances and an occasional "God bless you" from strangers. But there is no follow-through on these affirmations in the work world.

Like many working mothers, I became far more sharply aware of the economic perils of pregnancy and child-rearing when my friends started their own families. I was still child-free. I remember meeting a friend for dinner after a job interview. She was wearing a giant, gray, cable-knit sweater that hid her expectant form. "I am not telling them that I am about to enter my second trimester," she said of her potential future employers, pointing to the heavy knit, as she ordered cranberry juice at the bar. She got the job and revealed her status after she accepted it. Soon after, she began sensing her new boss's displeasure. At work, her thirty-six-year-old colleague was afraid to get pregnant for fear she would lose her job. And a few more expecting friends were covering up, literally and figuratively. Meeting with prospective employers, like law firms and news organizations, they cloaked their bumps under oversized jackets.

These fears were not unreasonable. Overall pregnancy-related discrimination charges are on the rise, increasing 23 percent from 2005 to 2011. These women felt like they had to "cover" the fact that they had what I call "hidden pregnancies." (When I mentioned this in a conversation with Nanau, she answered: "Women who hide their pregnancies are right to employ deceit. After all, how else can they negotiate with their employers?")

These negative stereotypes are part of the "caregiver penalty": a broad theory of what amounts to social punishment for being a caregiver. As the philosopher Martha Nussbaum points out, America penalizes the caring classes—mothers, fathers, day-care workers—and deems them "less than." This attitude stems in part from an intolerance for human weakness, and thus for those who serve humanity.

Parents face a specific variety of the caregiver penalty. The "parental penalty" is the workplace punishment for caring for their children. (Federal and local governments have fought like hell against parental leave laws.) The parental penalty is not just imposed on mothers: men also face it. Employers may slap down any fathers who wish to take a paternity leave. At one point, Nanau cited the research on this solemnly, as if it were source code: male workers on average get a boost in salary after they have children, unlike their female colleagues, but if they take paternity leave, they may pay for it professionally. As a Deloitte survey of adult workers found, one-third of male respondents said that they would not take paternity leave because taking time off would jeopardize their position. And they could be right. Jennifer Berdahl, an associate professor at the University of Toronto, has found that men who take on more of the caregiving of their children than is typical are more likely to be mistreated at work.

One might say: penalization of parents can know no gender.

Other elements of the parental penalty are both the scarcity and high cost of child care. This is a country that has not cared enough to support affordable and accessible child care for working parents—one of the biggest impediments for those either trying to sustain or aspiring to a middle-class life. The parental penalty percolates through the lives of those fighting to stay in or enter the middle class because it makes their lives that much harder and expensive and their schedules that much more maddening.

The motherhood penalty has shown that, as a subset of parents, we are punished the most. As Shelley Correll of Stanford University notes, employers are less likely to hire mothers. In addition, they are rated as less promotable and less likely to be suggested for management positions. Another study has found that mothers are hired for annual salaries an average of $11,000 less than the salaries offered to their equally qualified, but child-free, peers. The motherhood penalty is also a state of mind imposed on mother-workers who, shamed by rejection by the marketplace, internalize the societal disdain for caregiving, the prejudice that considers caring for children feeble or undistinguished and, even worse, not actual labor.

Women may be harassed when they are carrying a child or forced out of their jobs after a child's birth. When one parent must quit their job to stay home with the kids, it's usually the mother, or one must become an underachieving, harried worker easily stereotyped as "failed." To a good number of bosses, the tiny fingers of their employees' infants contaminate the proverbial pies of productivity. In the workplace, employers may see the breeding worker as an albatross in stretch pants. And pregnancy bias contributes to the squeeze on American parents by stymieing parents, particularly mothers, economically. The scholar Gillian Thomas has found that even if women aren't harassed for their pregnancies or pushed out of their jobs, their salaries go down—by 7 percent per child, according to Michelle Budig, a sociology professor at the University of Massachusetts Amherst.

This unsparing social attitude toward parental leave hits new mothers the hardest, since the pressure to return to work carries an especially heavy emotional and physical cost for new mothers. Having children is poison in lots of lines of work, evidence that you are not buying into the profession 110 percent.

It shouldn't have to be that way.

AS WE WALKED AROUND NANAU'S NEIGHBORHOOD OF GLENDALE, Queens, and the neighborhood she was raised in, Forest Hills, on a freezing day in 2016, we traversed seeming dioramas of upper-middle-class life as it was conducted forty years ago. There was an off-label French bakery, where she purchased an apple tart, and an old-fashioned tailor shop with a mannequin's torso, draped in a crepe de chine blouse with a bow, in the window. In Glendale, which was dominated by cemeteries and giant thoroughfares, Nanau had been able to buy a home for half the price of homes in wealthier areas of New York City, and some homes on the market were selling for even less. In her mini–Tudor house, Nanau's daughter Olivia, nearly nine, rolled on a giant red ball, did headstands, and rushed to show me a picture of her Elsa-from-*Frozen* Halloween costume. Nanau's seventy-five-year-old mother, Inga, a former accountant for nonprofits, was babysitting: she told me her own pregnancy discrimination story. One of her bosses in the early 1970s had addressed all questions about her employment while she was pregnant to her husband rather than to her.

Nanau was the daughter of German and Romanian parents who came to Queens in the 1960s. Her taskmaster father had made her study math obsessively as a child, and she had poured her resulting intensity into her legal practice. I was not surprised when she introduced me to the beloved pet she described as "the sweetest": an orange pit bull.

Nanau then recalled her resentment over feeling discriminated against when she was pregnant, which she had kept buried for almost a decade. Her memories of being mistreated had lingered even through the scrim of time. Nanau now had her own legal practice representing people—mostly women—who alleged that they had experienced discrimination in the workplace.

She also understood what it takes for a parent to care for a sick child, especially for mothers and fathers who lack money. When her daughter was born prematurely, Nanau stayed with the baby in the neonatal intensive care unit, where she was sent to be treated for jaundice and low weight. Nanau recalled the moment when she saw that the tiny sunglasses her daughter wore—to protect her eyes while she lay under heat lamps—had fallen off her daughter's face. Starting then, as if suddenly aware of her infant's fragility, she was afraid that her daughter would be damaged in some other way during the first year of her life. Nanau drank a gallon of milk a day to improve her breast-feeding, hoping it would make a difference for her underweight and premature girl. What if she hadn't been able to leave work to support her daughter's physical needs? Or what if she had been punished professionally for staying close to her undersized infant rather than her clients?

When Nanau built cases for other women, she looked for veiled references and common insults in their bosses' remarks that betrayed their gender stereotypes against her clients. Some of these remarks took the form of "helpful" hints that the employee shouldn't have had kids, or that she could stand to lose some pounds postpartum. In court, Nanau turned these comments into assets—the keys to showing the injustice and prejudice around childbearing that still infiltrates so many workplaces.

I spoke to some of these women, like a group of female airline pilots who were also moms had launched a case through the American Civil Liberties Union concerning their right to pump breast milk in the workplace. The claim was "that the company has discriminated against them and other female flight attendants by failing to provide accommodations related to pregnancy and breastfeeding." One of the pilots in the suit, First Officer Randi Freyer at Frontier Airlines, had been working at the airline since September 2013. She had her first child half a year later. After the birth, she was told that she couldn't pump on the job, which for her happened to be flying planes.

That's when she went all mama bear, as she puts it.

Up until then, the breast-feeding pilots and flight attendants "accommodated" themselves and pumped on the road. However, the airline "told us we couldn't pump on the aircraft at all, and no other options were given to us," according to Freyer. So they pumped in the flight bathrooms.

The pilots had asked for pumping stations on planes or in their airport areas, from Denver to Chicago to Florida, but were never given any such resources by the carrier. Eventually, after Freyer had her second daughter and encountered the same level of airline recalcitrance, she and a group of pilots sued on behalf of her and other employees like her who were told not to pump milk on the job.

Freyer wondered why the airline couldn't provide a clean place with an outlet for her mechanical pump. She would have even knuckled under and taken paid leave to breastfeed, but there was no paid leave to be taken. She had already been forced to stop working when she was thirty-three weeks pregnant, as women are not allowed to fly after that mark, and so she had already taken eight weeks of unpaid leave before the birth of each of her daughters, leaving her family short on funds. She would have worked a ground position, she said, if she had still gotten paid, but Frontier didn't offer her this option.

She lived in the mountains of Colorado in a town called Eagle. Fittingly, she and her husband are flight-mad pilots, he in the U.S. military and she for commercial airlines.

She was not a self-indulgent hipster mom with a breastfeeding-till-the-kid-is-three fetish. Freyer herself was surprised by her rebellion. She didn't like to stand out at work. She preferred, she said, to "keep my head down and do the best job I can do for the people who own this company." The battle with Frontier was painful: the tension was thick, she said. The Frontier managers who were so unhelpful had been her mentors, she told me. Her choice to pursue the lawsuit against the airline still struck her as "career suicide." But she knew she had to fight for it.

However, some discontented mothers often don't go to the courts like Freyer did. Instead, they take to that less powerful arena, the Internet, to vent—as they do on the pungently named website Pregnant Then Screwed. There women post accounts of how their pregnancies got them fired or otherwise warped their careers. A carpenter's apprentice was told that her job performance had been suffering since she got pregnant. A legal worker, Ivy League educated, was initially much praised by her supervisors, but her story also ended in "job elimination" once pregnant. It would seem that women's livelihoods are jeopardized by motherhood, starting at conception. Why? As President Donald Trump once commented, pregnancy is "certainly an inconvenience for a business. And whether people want to say that or not, the fact is it is an inconvenience for a person that is running a business." However loathsome, he was articulating the cruel *common sense* of capitalism: why should employees take any kind of leave for any reason at all, least of all for reproduction? But by this logic, what about those of us who do reproduce?

I discovered all this when I finally got pregnant. I read the novelist Rachel Cusk's cool-eyed memoir of her gestation and early maternity, *A Life's Work,* and folded down the page where she writes: "after a child is born the lives of its mother and father diverge, so that where before they were living in a state of some equality, now they exist in a sort of feudal relation to each other." After the births of her children, she says, her slide "into deeper patriarchy was inevitable." I felt that slide too, but in a way that was as much about the "patriarchy" and the feudal nature of the world of American work as it was about the gendered state of my marriage.

It was feudal that, as the would-be mother, I had already become a serf, an unpaid servant carrying my fetus as if it were a princess. I had resisted having a child, not wanting to be indentured in such a way. And I didn't even know just how bad it would get. Like Nanau, I was middle-class, and at the time of my pregnancy I was self-employed, part of the Middle Precariat. I was thirty-eight and living, as I do now, in one of the most expensive cities in the world, New York, with my husband. The promise that I would be ailing for only the first trimester was issued with calm, even smug, smiles by my ob/gyn and other mothers. Yet at five months along, I was still so sick that I felt like a fainting Victorian woman. I could only keep down mango smoothies, but their cost seemed prohibitive. My particular breeding symptom was hyperemesis, the Latin word for regurgitating the taste of prenatal vitamins! I regularly found myself retching into metal wastebaskets on street corners. In those months, every street, office, or store smelled profane. For me, pregnancy was a bad drug that turned me puce with ailments. My friends proffered lemon pregnancy pops and gray anti-nausea bracelets, which seemed comically feeble, woo-woo tools to push back overwhelming waves of bile. I couldn't work or even manage to have a conversation. I applied the name of the mainstream chain store Destination Maternity to myself in grave philosophical fashion: I was frail now and felt like I was heading toward a place that would be as much a terminus as a blazing new beginning. Given how ill I was, I couldn't work as a freelancer, as I had done for fifteen years. So we lived off earnings from my husband's assignments and our savings.

As I lay in bed reading, I knew our future family was growing poorer and poorer, but all I could do was stay under the duvet, my stomach enormous, tensile, and blue-veined. I became a morbid insomniac reading about nineteenth-century women's afflictions and watching late-night promotional TV; the commercials for knives and thigh exercisers seemed pornographic to me and underlined how ineffective I felt.

When my daughter was born—a pellucid alien, her head sculpted into a Mohawk of flesh by her birth—we were already racking up unexpected bills. Though we had purchased freelancers' health coverage, we still had substantial hospital costs left uncovered. According to a study written in 2013, the cost of delivery, for both vaginal and cesarean births, has nearly tripled since 1996: the average cost for a C-section in the United States in 2013 was $16,038, compared with $12,560 for a vaginal birth. Kids don't come cheap.

I remember how nervous I was after her birth when my bank account numbers fell: I was paying for my own maternity leave. And I was a miser to myself, celebrating every hand-me-down onesie. I remember breast-feeding while recognizing that my bank account was being eviscerated, feeling an anxious pinprick when my baby latched on. I never produced enough milk, so I pumped in an attempt to make more. The pump sounded like electronic music or a disco, an industrial lullaby, and I pumped in offices and train bathrooms as I went to various journalism and editorial hustles. The need to earn money another constant sound, if less audible, in my head. Acquaintances asked, "Are you planning on another?" I knew my daughter was destined to be an only. With just a scrap of maternity or paternity leave, many parents choose to have just one child, as my husband and I did, because it's all they can manage economically. Our family was far from alone. Only 14 percent of American workers have paid family leave.

And that's part of the reason why many American women can't afford simply to exit stage left from the workforce for a few months when we start to breed, then return to our jobs. There is always the threat that the precarious, biology-denying market might annihilate us.

One needn't travel far to see a different kind of workplace. Women in France, Britain, Chile, the Netherlands, and South Africa all pay less for vaginal deliveries in hospitals, and much or all of that cost is covered by insurance or the state. In the Netherlands in 2012, the average cost for a conventional delivery was $2,669, and in 2015 that cost averaged $2,000 in South Africa.

And on to family leave. In Quebec, Canada, for instance, the state maternity and paternity leave insurance gives a parent up to 75 percent of his or her salary and covers up to either fifteen or eighteen continuous weeks for the mother and three to five weeks for the father. In Denmark, mothers get four weeks of paid "pregnancy leave" *before* their child is born, in addition to fourteen weeks of paid maternity leave after giving birth. And parents who are adopting can take time off as well—fourteen weeks. Parents may receive up to fifty-two weeks of paid leave per child from the government in total, including those first fourteen weeks. Before I had my daughter, I had reported in Iceland. At one tourist stop, a female guide showed us around the carefully restored mid-century house of the country's most famous author, Halldór Laxness. The guide was quick to describe her life—that she had three children and no current partner, yet her family was thriving. (Most likely, I speculated, all of them were cheerfully soaking in hot springs together.) If she'd had a husband, he would have probably benefited from Iceland's "daddy leave," which 90 percent of its paternal population receives. The country provides three months of parental leave for both mothers and fathers, and then three extra months of leave to be shared as they see fit.

It's worth noting that that nation's economy, since that spot of banking trouble almost ten years ago, is doing just fine for the moment. Of course, the market—which doesn't value reproduction—still structures lives in all of these enlightened countries. In Sweden, fathers do not take parental leave to its full extent because they see it as a risk to their status and competitiveness at work, so the government has implemented incentives to make sure dads take the leave they are entitled to. While we might mock the sort of Euro-chauvinism on display in books like Mireille Guiliano's *French Women Don't Get Fat,* Pamela Druckerman's 2016 *New York Times* essay on living in France as an American was indeed envy-inducing. "Suddenly it wasn't all on me," Druckerman wrote. "I gradually understood why European mothers aren't in perpetual panic about their work-life balance, and don't write books about how executive moms should just try harder. Their governments are helping them, and doing it competently." Statistically, middle-class parents have it even better in Finland—and they also fare better in, say, the Netherlands (with none of the Finns' arctic unhappiness).

After my daughter was born, I fantasized about long paid weeks abroad—perhaps in Copenhagen, where I'd nurse my bonny infant on a Hans Wegner chair while watching the Danish political TV show *Borgen*. My husband would be serving me pickled fish and black bread. In contrast, for me and so many other American parents, it can seem as if we are being asked to work jobs we hate to support families we rarely see and therefore don't really know. In the United States, only 13 percent of private-sector workers—men and women alike—get paid family leave through their employer, though there is scarce data on the duration and quality of that leave. Some of the best American companies in terms of leave are tech companies or platforms (Adobe, Spotify, Etsy) and nonprofit foundations. The rest of us are being squeezed like fruit, until there is only rind left.

That feeling of being wrung dry by child-care costs and lack of leave partly explains why reading about raising children in France has become a kind of porn for U.S. working moms. It's "bringing up *bébé*," sure, but it's also "bringing up *das Baby*" in Germany or "bringing up *il bambino*" in Italy (where women get five months' maternity leave at 80 percent pay). In many countries, paid leave can extend to three months or more for a new parent. For new parents, the United Kingdom provides 280 days at 90 percent pay, Canada 119 days at 55 percent pay, and the Netherlands 112 days at 100 percent pay. Anyone who has had an infant knows that the first three years are not only the most demanding and time-consuming time but also the most critical for development. (The perfect audio for this discussion would be the sounds of crying babies turning into contented goo-goos once they realize that their mothers can stay home with them at nearly full pay.)

Nationally, the Family and Medical Leave Act (FMLA) grants up to twelve weeks of *unpaid* leave. FMLA applies to public agencies, elementary and high schools, and companies with fifty or more employees. The population not covered by it will only grow, however, along with the rise of the Middle Precariat. After all, the middle class is increasingly made up of freelancers and contract workers. It was fitting when the unofficial paper of the American middle class, the *New York Times,* noted that our country's bourgeoisie, long the most affluent in the world, no longer holds that status. The Americans in this book are not all that ritzy: as mentioned earlier, I use the U.S. Commerce Department's definition of the middle class, which includes aspirations as well as income—

aspirations such as owning a home or having a car for every adult in the family.

THE MOUNTING PRESSURES OF PREGNANCY AND EARLY MATERNITY as well as the bias that pregnant women may experience stem from villains beyond the alpha male bosses or porcine pols in our lives. At root, like so many of the things that have created the pressured parents of today, the rise in pregnancy prejudice reflects our country's denial of women's—and sometimes human—biology. We are routinely precluded from having access to birth control and birth control coverage, as if the causes of unwanted pregnancies were still a mystery. We are given—two places—for breast-feeding in public places. And as I explore later in the book, older people's biological age is routinely held against them in the workplace: they are made to feel like failures if they cannot "overcome" their biological years and if their résumés give them away.

Of course, the question of whether our biology defines our lives and careers has a complex history. At one juncture in the fight for equality, feminists shouted, "Biology is not destiny." But that mantra did not mean that biology doesn't exist at all. It is certainly present when women are considered valued, productive employees when they behave like men—or better still. Nanau recalled how she felt working later on in her pregnancy at a different firm—how she tried to blur her gender by "dressing like a little man" in an all-black pantsuit, her thick brown hair cut close to her skull and gelled. Being pregnant, though, she felt unprotected, and so she faux-swaggered down the street, she said, taking up as much space as possible.

This seemed to me another kind of hidden pregnancy: both Nanau's swagger and other women's attempts to hide their pregnancies from their employers for as long as possible are what the great sociologist Erving Goffman would call "covering." These are examples of the ways people mask and obscure identities that are stigmatized, or what Goffman calls "spoiled identities." And being a pregnant woman in America, no matter how many reality shows and greeting cards say the opposite, *is* a stigmatized identity.

This shows up in how female middle-class workers, like Nanau, can be punished for having a baby too early. Postpone, plan, and wait are the watchwords of a middle-class professional. When one's employer determines one's family planning, everything is upside down.

If you add institutionalized racism and America's race-based wealth disparities to the gestation equation, you will swiftly discover that the median black family has it even worse. That mom has far less savings to draw upon than their white counterparts for things like pregnancy or maternity leave, or in the case of costly medical bills during pregnancy or after childbirth. (The median wealth of black families can be found in the 2017 report "The Road to Zero Wealth: How the Racial Wealth Divide Is Hollowing Out America's Middle Class.")

These are the battles that Nanau's mother—and my own mother—had assumed they had won. The nearly forty-year-old Pregnancy Discrimination Act (PDA), which amended the Civil Rights Act of 1964, asserts that discrimination on the basis of pregnancy or childbirth is unlawful sex discrimination. An employer can't refuse to hire a woman because of a pregnancy-related condition alone. Congress approved PDA in 1978, when I was six. "PDA brought about some immediate and significant changes in employer policies," wrote the legal scholars Deborah L. Brake and Joanna L. Grossman. PDA sought to untangle all the crappy and gendered bias by arguing that a pregnancy shouldn't affect a worker's pay, promotion, or termination. Over time, however, and thanks to interpretations by the lower courts, the PDA's scope shrank and its idea of pregnancy discrimination narrowed. This legislation got only part of the way there because it offered a limited vocabulary for describing discrimination itself. As a lawyer, Nanau herself had seen that "employers know how to build paper trails" around female workers who complained; she had seen how they used these paper trails to defeat women's cases. This is part of what the legal philosopher Joan Williams calls the "maternal wall," a spatial metaphor that, when attached to the proverbial "glass ceiling," virtually imprisons female workers. The pregnant worker, in other words, is subject to several different biases that stem, at least partly, from the disregard for our private lives shown by our workplaces and, at times, our employers. The result: working mothers are regularly asked to cleave themselves, separating the leaky processes of birth and early child-rearing from their careers.

What changes might protect the squeezed female worker? Groups like the Gender Equality Law Center (GELC) have some ideas. They want "pay transparency" for both men and women and gender-neutral paid leave laws. The GELC encourages men, even those who are not "high-level" workers, to take—or just ask for—paternity leave; the group also applauds the best corporate child leave practices and calls out the worst. Other nonprofits are following suit, and there are also major lawsuits pending on behalf of pregnant women and new mothers in the workplace.

If all of America's businesses supplied parents with nine months of family leave, fewer managers and bosses would have to bear the pressure of workers' pregnancies: all companies would suffer alike. If the pressure were widespread, the costs associated with leave would most likely become more acceptable.

Some have argued that discrimination itself costs us billions of dollars in diminished output, job turnover, and the time, effort, and money to file discrimination claims. There is also the price paid by the employer with the reputation as a corporation that sucks the life out of employees and shuts out pregnant women. This can sully a public image, as it has for technology companies that were once lionized as labor utopias and then called out for being high-end sweatshops. GELC wants companies, small businesses especially, to understand that, as the organization's staff lawyer Lauren Betters put it, they won't go under if women take time off.

Yes, pregnant workers may be less profitable employees than women who aren't gestating. And that's okay. We can simply choose to value aspects of life beyond economic productivity. Women also shouldn't have to suffer when they are honest in their workplaces.

"I'm a complainer, and I believe in complaining when you are mistreated," Nanau told me. "But it's going to cost women when they come forward," she admitted.

In late October 2016, Nanau showed me around her home, furnished almost entirely in the collectibles of her and her husband's forbears. As we sat among the mod pillows, it was hard to tell what decade we were in while we drank herbal tea and ate shaved carrot salad in her sunken living room, occasionally checking the time on her grandfather clock. She pointed out that the photos lining the wall near the staircase showed generations of workingwomen in their families, in both the United States and Germany. "We weren't born with a ton of money," Nanau said of her family and her husband's family. "We might have gotten fancy jobs, but we had to keep working at them. Middle-class people,

people like me, can't take off a couple years from work during and after they are pregnant."

She put off pregnancy, thinking that when she finally "had children, it would be different and not only the wealthy could afford to have kids." Saying that, Nanau seemed nostalgic for the idea of the future she had in the past.

It's not just a problem of her personal history, though.

Her story today and others like hers are just part of something much larger.

In the ensuing decades, the changes in how America addresses and supports families—and maternal workers—never transpired. This is the "class ceiling" for mothers who are not among the wealthiest, a toxic mix of gender and economic status.

These limits still shape—and distort—the life for many of us.

2

HYPER-EDUCATED AND POOR

Professor Bolin, or Brianne, as she told her students to call her, might as well be invisible. When I arrived at the building at Columbia College in Chicago where she taught composition, I asked the assistant at the front desk how to locate her. "Bolin?" she asked, sounding puzzled, as she scanned the faculty list. "I'm sorry, I don't see that name." There was no Brianne Bolin to be found, even though she had taught four classes a year there for years. She didn't even have a phone extension to her name, never mind an office.

Bolin rushed in late to the lobby—she'd offered to give me a tour. Her red hair was pulled back in a ponytail, and red electrical tape was wrapped around the left temple of her black geek-chic glasses; they had broken a few months before, and she couldn't afford a new pair. Bolin dressed up for the occasion: a black vest (from a thrift store, she'd tell me later), jeans (also thrift), and a brass anatomical version of a heart dangling at her throat from a thin black string. This was a rare and coveted evening off for Bolin, the mother of a disabled eight-year-old boy named Finn. Finn's father's fiancée had agreed to babysit for her, but so far she was too agitated to enjoy the time off. She had just learned that the woman and Finn's father, a blacksmith, were getting married in a few weeks, and they wouldn't be able to take care of the boy during that time. It would all be on her, again.

After she showed me the computer lab and some of the students' abstract photography and video installations, we settled in to talk in the student lounge, which featured sleek modern furniture and high-rent views of Grant Park and Lake Michigan. By this time, Bolin seemed more angry than anxious. An adjunct professor, she earned $4,350 a class, and never more than $24,000 a year. At the moment, she had $55 in the bank and $3,000 in credit card debt. She was a month behind on the $975 rent she paid for a two-bedroom house next to railroad tracks in a western Chicago suburb, where every twenty minutes a train screeched by. Her bookshelves were full of poetry and philosophy from grad school that she could recite from memory, and she collected French 1960s LPs, but she had to rely on food stamps to feed herself and her son. And because her job didn't offer health insurance, they were both enrolled in Medicaid, the state and federal health-care program for the poor. Coverage for a child Finn's age in Illinois was capped at an income equaling 142 percent of the federal poverty level—about $22,336 in 2014 and $23,060 in 2017—and Bolin couldn't make more than that.

It wasn't supposed to be this way. Bolin, the English major, knew that was a cliché, but she couldn't help thinking it all the time. *It wasn't supposed to be this way.* In college at Eastern Illinois University downstate, she had inhaled books—lived "in a trailer park with a friend, reading the novels of Virginia Woolf and Marguerite Duras, getting into Kerouac and Ginsberg and that Beat rebellion thing," she recalled. She had earned a bachelor's and a master's, studying avant-garde poetry. She didn't expect to become an academic star—Eastern Illinois wasn't the University of Chicago—but she did assume that she'd have a steady job with adequate pay. "I like nice things—I'm a little bourgeois," she said. "I thought, at thirty-five, I'd have clothes without holes in them and money in the bank, but I shop at Goodwill exclusively. I wear Banana Republic $5 suit jackets that wear out quickly because they've already been worn so much beforehand. My dreams did this to me. It's not a shameful thing, although I wonder if there *is* something wrong with me."

Much political rhetoric these days is devoted to the importance of broadening access to college—and there *is* plenty of evidence that a degree improves financial prospects—but in the post-crash world of today, a good education may not keep you from hovering near the poverty line. The number of people with graduate degrees receiving food assistance or other forms of federal aid nearly tripled between 2007 and 2010, and those with a Ph.D. who received assistance rose from 9,776 to 33,655. More specifically, at least 28 percent of households that used food stamps in 2013 were headed by a person with at least some college education. This proportion was 8 percent in 1980, according to an analysis by University of Kentucky economists.

This is part of the squeeze that affects parents and their children. The hyper-educated poor lurk alongside the women who are victims of pregnancy discrimination. If American parents with advanced degrees can barely pay for their children, what does that mean for other American parents? How can a pregnant worker fight back against lack of parental leave or bad workplace accommodation when she knows how few options there may be for her in the professional world? How can a worker whose day-care expenses are out of line with her wages feel as if she can complain? After all, she knows how hard it is to get—and to keep—a job at all. A survey of five hundred adjuncts, for instance, found that 62 percent of them make less than $20,000 a year from teaching. If they complain, they may simply not be assigned the classes they need to get by.

The pregnancy discrimination described in the last chapter is one way in which our very bodies are squeezed by the work world. The hyper-educated poor described in this chapter are a prime example of a different but related thing. In today's America, one may be unlikely to "pass down" cultural and social class standing to one's children. After all, will children, once they reach college, be able to afford the educations their parents had? Or—egads—graduate school? And will they retain their parents' social confidence—their belief in their professional narrative? Will they even have a career arc at all?

The hyper-educated poor are as hidden to the country at large as Bolin is at the college where she taught. "Nobody knows or cares that I have a Ph.D., living in the trailer park," said a former linguistics adjunct named Petra, the mother of one child, who lives in Eugene, Oregon, and was on welfare and food stamps. Michelle Belmont, the Minnesota librarian and web developer who admitted that few of her friends had any clue how broke she was, put it this way: "Every American thinks they're a temporarily embarrassed millionaire: I am no exception."

These professors and other extensively trained and educated workers have all the typical problems of the Middle Precariat: debt, overwork, isolation, and shame about their lack of money. They also may have very little time for leisure, not even for a few dates over pale ale with their partners or get-togethers with friends where they can confess their woes or snicker over gossip. They take almost no holidays.

Many of them told me that their parents were more economically comfortable than they were, even though their parents often had far fewer educational attainments. Whenever I talked to these Middle Precariat parents, I also heard the ring of self-blame and ridicule. Was it a sin to have pursued a high-minded profession and to want nice things? They felt like it was.

Their lives were also unlike the more cushioned lives of their older colleagues and certainly did not resemble the trajectories they were expected to follow.

Bolin herself kept in touch online with a large circle of her fellow travelers, including Justin Thomas, a friend from college with a master's degree in history. An adjunct at Parkland College and Lake Land College, about three hours south of Chicago, Thomas taught between four and six classes a semester and in 2017 earned $1,675 per class at Lake Land and $3,100 per class at Parkland. His paychecks arrived a month after each semester began, he said, and during those four weeks it was macaroni and cheese and baked potatoes every night for his two daughters. (Because he didn't have full custody of his children, he wasn't eligible for food stamps.) "I say, 'Sorry I can't afford to buy you anything, even an ice cream,'" he said, getting choked up as he added, "For me to help my daughters with their dreams, I have to give up my dreams." Though he had been moonlighting for his father working in construction, money remained tight. "I'd love to get my daughter music lessons—she's talented. But right now I don't have the resources to take advantage of her ability."

It's not just academics who are highly educated and downwardly mobile. Other respected professions are losing their luster as well. Employment for recent law school graduates went from almost 92 percent in 2007 to 84.7 percent in 2012 to 87.5 percent in 2016, according to the National Association for Law Placement, and the average law student's debt was about $140,000 in 2012—a 59 percent increase over 2004. (You will meet some of these sufferers in later chapters.) Other professions that haven't regained many of the jobs lost during the recession include architecture, market research, data processing, book publishing, human resources, and finance—all of which either require or tend to attract workers with master's degrees. Making this squeeze worse, to my mind, was the oft-heard mantra "do what you love"—the exhortation to members of the middle class that they should try to make a living pursuing their *dreams*. Well-meaning mentors and corporations recited this dictum. I myself have heard it often. Those exhorting others to "do what you love" manage both to look cool and to squeeze more labor out of their workers. This advice is meant to correct old ideas that labor should be dutiful and even servile rather than impassioned. Labor has been increasingly construed as vital to one's character and sense of meaning and value. It wasn't always thus, as some historians have argued. Work in previous centuries was something people did mostly out of economic urgency, and they dreamed of a heaven where they wouldn't have to work or a land of milk and honey. (Aristocrats, of course, tended not to work at all and to live off their property.)

To bring this mantra abruptly into the present; I had typed this sentence at a desk at the coworking behemoth WeWork, where the phrase was emblazoned on the company's T-shirt. I stared at this koan, hoping the young woman wearing the shirt while restocking WeWork's coffee was living it out somewhere else, and I thought about how many parents I knew, both younger and middle-aged ones, had bought into this maxim, to a mixed end. How are we all supposed to survive doing what we love? What will become of the rest of the world? The exhortation can even feel coercive, when applied to the lot of journalists, for instance, or start-up tech workers and other employees in floundering but "creative" businesses. They labor long hours for low salaries doing jobs whose glamour or promise of renown is meant to be their own reward.

Americans may well be marinating in an atmosphere of "cruel optimism," as the scholar Lauren Berlant wrote in her book of the same name, an atmosphere that "exists when something you desire is actually an obstacle to your flourishing." Among the impossible objectives embodied by the directive "do what you love" is the fantasy of the good life. Individual security and promise have frayed, wrote the University of Chicago professor Berlant, including "upward mobility, job security, political and social equality, and lively, durable intimacy," and yet we still believe we can "have it all" even when America no longer aids us in having our lives and careers "add up to something." What creative-class members like myself have experienced by "doing what we love" could sometimes be likened to Berlant's "stupid optimism," which "is the most disappointing thing of all."

As Miya Tokumitsu wrote in her book *Do What You Love: And Other Lies about Success and Happiness,* "doing what you love" in America has been so co-opted by corporations to better exploit their workers that it's a husk. Examples of this abound, from the technology companies to upmarket restaurant chains and stores like Trader Joe's that all insist their employees are—or act—happy. There are also, of course, more subtle types of exploitation, as the adjuncts and schoolteachers and journalists who have gone down the virtual garden path of "loving their work" well know. They might, in time, find themselves forced by their professions to work for almost nothing, all in the name of pursuing their avocation. There is something uncomfortably classist in the insistence on doing what you love in the first place, I realized after putting my own "creativity" aside to edit for hire at various points in my life. If the insistence on "doing what you love" comes from a place of privilege, where risks are lower and failure is not the be-all and the end-all, what happens to those who don't have that privilege?

During our current age, what I sometimes think of as "The End of the Middle," to choose to have children further rocks an unstable structure: we are playing an existential game of Jenga.

The precarious workers like Bolin and others were educated to be highbrow. They drank from the chalice. If you find that you are indeed detached from your vocation, the question arises: *who are you?*

The desire not to be alienated from her work started early for Bolin. She felt herself to be different from the other kids at her school in small-town central Illinois. For starters, she was adopted and an only child. Early on, she tested into a gifted program and was the kind of smart that her mother proudly called "scary," though her mother's assessment morphed into "troubled" when Bolin became a teenager. In high school, she didn't fit in—she was very emotional, wore all black, and read constantly.

At college, however, she quickly felt like she'd found her place in the world. "Literature gave life an extra resonance," she said. Then, in her sophomore year, Bolin's boyfriend was, shockingly, murdered by his roommate. The trauma led her to dive yet more deeply into literature, especially poets such as William Carlos Williams and George Oppen. *What is the soul?* she wondered. Was her boyfriend's soul out there somewhere? "I had camaraderie in a personal way with the authors I studied," she said. "I was a lonely person—I still am—and books made the world a more beautiful place."

No one at her college mentioned that becoming an academic might not be the wisest career path, she said. Instead, her favorite professor, Michael Loudon, who taught American Romanticism, encouraged her to come to his office and sit and talk. (He is now retired.) "He had faith in me: he knew I'd continue with the ideas I was working with and write a dissertation. No, he didn't think I'd have a big career, but he was sure I'd have decent work. That was a given."

In 1975, during Loudon's collegiate era, full-time tenure-stream professors were 45.1 percent of America's professoriate. As of 2011, they are only 24.1 percent: only one professor in six (16.7 percent) actually has tenure. Adjuncts or part-timers like Bolin are contingent, meaning that they are non-tenure-track: many of these exquisitely trained and educated people are financially on the rocks. This change in academia had begun by the time Bolin went to college, but neither she nor her parents were aware of it. Her father, who hadn't gone to college, built tires for Firestone; her mother was a homemaker who had an undergraduate degree in home economics. "Clocking in at nine and home by dinnertime," Bolin said of her dad. He worked to live, not the other way around, and he didn't necessarily understand his daughter's quest for work she *loved*.

Nevertheless, Bolin's parents, who paid for her undergraduate education with savings, were impressed when, in her midtwenties, she graduated and immediately started teaching composition at Westwood College in Chicago. (After one semester, she switched to Columbia, the Chicago art college.) She'd hoped to teach literature, but she came to love basic writing and comp, she said, and enjoyed helping her students learn to write cogently. Also, Chicago thrilled her. She'd never seen people of so many races and nationalities. She could hear eclectic music night after night, and especially loved klezmer and Balkan folk music. She even formed her own two-piece band, Mud Show, with a cast of instruments such as an accordion, a bass made from a steamer trunk, a bucket of chains, and a typewriter. "I was making it financially, but I was living as a twenty-six-year-old in the city, with several roommates, hosting house parties with live music, enjoying life," she said. "I had no serious partner and no future plans. I was living an extended youth."

Then, at twenty-eight, she got pregnant, the result of a random hookup with a twenty-year-old in a band she liked. She knew she'd be raising her child mostly on her own, but Bolin said she never considered not having the baby. Then further misfortune struck: Finn was born with quadriplegic cerebral palsy. To devote herself to his care, she quit work for several years and moved back in with her parents. Her mother remained proud of her daughter, no longer as an academic but as a caretaker of a boy with bright blue eyes and a crown of sandy-brown hair who couldn't eat or walk without assistance, whose ribbon-thin body had to be lifted in and out of his wheelchair many times a day.

In 2008, when Finn was two, Bolin returned to Chicago and started teaching as many classes as she could get from Columbia, but her boss warned her that she'd never get a permanent job. "Academia just isn't a career choice anymore," Bolin said. There was little support to be had from the people who criticized her for getting "knocked up," who blamed her for getting herself into this fix by getting pregnant "out of wedlock." These nasty naysayers were irrelevant to Bolin, however: their remarks, though hurtful, couldn't change her trajectory. Her situation was what it was. She belonged to a class of people who suddenly find themselves with neither extensive social support, besides parents and an online social network, nor a clearly delineated future.

One reaction to this plight might be: *Get over yourself! Find work that pays the bills!* Or, as Karen Kelsky, a former anthropology professor who founded a counseling service called The Professor Is In, has put it: "find a 'real job.'" (I paused at this, wondering how being a professor was not "real" in Kelsky's mind.) Kelsky's clients pay her $150 per hour to edit their documents or résumés. Two email consultations are $15. An hour-long Skype session for interview prep is $250. This is all part of the guidance she offers clients in reinventing themselves, she said, and, for some, in expressing "rage, despair, and disappointment" about their disappearing profession.

"Adjuncts can accrue massive debt to support their children," she noted, and in the process "destroy their health, teach at five campuses, in a professional death spiral. Once you've given it your best shot, it's time to move on." She helps people with postgraduate degrees identify other marketable skills, such as analysis, data gathering, writing, and public speaking, when they can't get on the tenure track or, having not gotten tenure after being on track, have to start all over again. The very fact that hired consultants and experts like Kelsky exist—in academia and in law and in other upmarket professions—demonstrates how troubled these professions are.

Even though Bolin is an avid follower of Kelsky's blog (she can't afford her one-on-one service, needless to say) and Kelsky's advice sounds sensible, Bolin, already stretched thin by working and caring for her son, said that she struggled to find time to send out her résumé or get additional training—the latter not being free, of course. She had thought about supplementing her income with some kind of retail job, but Finn's child-care costs would eat up her paycheck. She had started to train as a speech-language pathologist a few years earlier—her son had needed speech therapy since birth—but the further along she got in her studies the more despondent she became, she said, and she eventually dropped out: her experiences with her own son were traumatic enough without having to consider other kids with similar struggles. In 2015, she had been looking into work as a campus union organizer, to capitalize on her interest in improving adjuncts' lot, but that hadn't really panned out either.

Bolin's situation was not just the result of too few hours in the day. As social psychologists who study what's known as "decision fatigue" have found, being poor takes a huge amount of mental work. There is a constant need to weigh the merits of spending even the smallest amounts of money: yes, maybe I should buy a few extra bars of that seriously marked-down soap (one of the experimental conditions tested by a Princeton economist in poor Indian villages), but then I can't afford this week's medicine or food. Tagging along with Bolin at Trader Joe's, whose very upscale-ness pointed to Bolin's fractured identity, I saw how tiring it was for her to try to stay within her monthly $349 food-stamp budget, for which she qualified only in the summer, when school wasn't in session. (She also could apply for about $600 in Supplemental Security Income [SSI] benefits in months when she earned less than $2,000.) Because Finn had to have expensive almond or rice milk—he was lactose-intolerant—Bolin would hunt for the 59-cent-per-pound giant bag of chicken legs and the 49-cent bag of carrots, and she never bought anything other than the absolute cheapest ground beef. "I read blogs about people wasting $20 on frivolous things like a photo booth or fancy cheese," she said. "I'll never do that."

When so much mental activity is devoted to basic survival, little is left to engage in long-term thinking or to muster willpower—as Bolin well knew. "I need to smoke to relieve the pressure," she told me as she feverishly rolled her own cigarettes one evening when I took her out to a bar, where she also found relief in the form of plentiful margaritas. She was self-medicating, she said; other times, she used Xanax for anxiety. She also took a daily antidepressant. As Linda Tirado, whose raw and honest blog post on her own minimum-wage existence catapulted her into the national spotlight in 2015, bluntly writes in her book *Hand to Mouth:* "Being poor while working hard is fucking crushing."

More broadly, education used to be a gateway to middle-class life, but now it simply doesn't get you there. Jobs in the academy and elsewhere are either endangered or moribund. Bolin is just one among many, fighting hard to stay in a class that may be melting away.

Bolin's desperation came through perhaps most poignantly when she took me to her favorite Chicago neighborhood, Andersonville. Her education had influenced her tastes, and she peered longingly into shop windows filled with midcentury antiques, wax flowers, handmade hats, toffee made with European beer. She told me that she was a foodie and loved pasta with cream sauce and shrimp cocktail, as well as the "opera cake" sold at the tourist-ready Austrian bakery, but that the cafés and restaurants in Andersonville were as out of her reach as the rents, which were around double those in her own Brookfield neighborhood. She stopped in a feminist bookstore, wishing she could spring for a book on sex and feminism, or a new collection of essays on living in the Internet age.

The next day Bolin and I spent the afternoon pushing Finn in his wheelchair through Chicago's Lincoln Park. At a playground, Finn rushed down the slide on his stomach, with a big smile on his face, but at other times he screamed in frustration at his physical limitations—he just wanted to run or kick a ball. What Bolin wanted shouldn't be so hard to achieve: an income of $35,000 a year in a steady version of the job she had. "Finn would be okay with that," she said simply. She'd be able to afford the next size up in a tricycle for him—though he could barely walk, he loved to ride. Bolin also said that she hoped to find a partner to love and share practical tasks with, such as child care, but she hadn't met anyone who seemed right for a serious relationship.

With twilight approaching, she pointed out an attractive dark-haired couple sitting with their toddler son on a bench, a fancy stroller parked next to them. "When I see couples who have jobs," she said, "couples who look perfect, I want to ask them: 'How did you do it?'"

Like all of us, Bolin was utterly unique but also utterly general. She was one of a layer of America's middle class, its brain-workers. She was also one of the many adjuncts who are further squeezed by being mothers. As a mother of a disabled child, she was pinched all the more. Bolin was also one of the workers who believe—rightly—that they are being exploited, the new disgruntled who are voting for independent or renegade presidential candidates. In 2016, she looked at each political candidate in terms of how much they would help the disabled, especially the poor and disabled, like her son. She didn't see much of a difference between them on that score.

Adjuncts like her are in an especially weak position. In many colleges and universities, part-time workers teach a tremendous number of classes, a fact that points right to the core irony of their position in our great teaching institutions: successful academic careers are built on the avoidance of teaching. Graduate students and part-time instructors shoulder a large proportion of the teaching, especially in introductory courses, and one of the perks of being tenured is a lower course load. Professors like Bolin are in a line of work created by the supply side. We must reconsider the long and deeply held belief that a graduate degree in a stable and ostensibly sensible field is the path to personal betterment. That is no longer a given route to success.

What does flourish in the academic marketplace is self-blame: the school-fee-indebted may castigate themselves for going to these places and getting the degrees, or for not turning their credentials into well-paid careers. The ideal, of course, would be for them to not blame themselves (or even worse, blame an oppressed minority), but instead to have a dimensional understanding of the economic and social forces shaping their experience—in other words, to know that their dilemma does not represent a personal failure but a system failure.

Brain-workers are also squeezed by another development: the withering of humanities programs as the priority shifts toward technical education. Industry and business are pressing for better-prepared applicants. After all, ours is a world in which a company like Google requires that many of their new hires have engineering degrees. And politically, countries want to stay competitive internationally through their prowess in technical fields. The social impact has been brutal. Whole sectors of employment are dwindling away, and many people who are untrained for the new economic reality find themselves stranded, not only jockeying to stay where they are socially but also at odds, personally, socially, and politically, with what they invidiously call "the elite." The spectacular rise of Donald Trump can be understood as an expression of the gulf between middle-class citizens and America's ruling classes. It should not have been news to anyone, since these changes have been under way since at least the Reagan era. Yet colleges and universities have been slow to adapt to these changes, which are already more than a generation old.

There was once prestige in having a humanities degree, but much of that esteem is now gone; this credential may now be seen as an antiquated honor. University culture has been slow to acknowledge this change, even as English departments collapse. Educational institutions have been moving toward the sciences aggressively over the past decade. Even university administrations may not support humanities programs anymore, citing the lack of money they accrue, or their low enrollments. The university culture at large is also shifting to vocational attitudes toward education.

All of these changes left Bolin in a fix. She came along in a moment when having earned the leisure and training to enjoy "high culture," for example, had shifted. Humanities programs may be shambling toward the sciences. At Yale University, Lisa Zunshine, now a literature scholar at the University of Kentucky, studied readers of modernist authors using fMRI technology. Rather than pursuing research in a cramped office or library carrel, she did her work in a lab. Bolin, being intelligent, could have retrained with such lines of inquiry in mind. But she had neither the time nor the money to get retrained when I met her. Her dilemma was not that she was stubbornly trying to stay put. It was that she was trapped and slipping.

When I was speaking with Bolin and professors like her, I was struck by how close I was to being one of these adjunct professor mothers myself. At first I had taken the same route to becoming an English professor, and perhaps also a permanent adjunct like Bolin, with a daughter I would struggle to support. When I was twenty-three and self-effacing, my hair in a tight bun, I read the lines "Obsessed, bewildered by the shipwreck of the singular," out of George Oppen's great 1967 book *Of Being Numerous*. I read this book on my hour or so commute on the Q train, headed to teach Brighton Beach community college students who had failed their basic reading and writing exams. During that season—in which one of my students left pieces of paper with obscene phrases on my desk—I tried to teach poetry to him and the others. I sometimes succeeded. I was a narrow and concentrated person then, a grad student hoping to start a "career" in poetics, of all ridiculous ambitions. It may sound absurd, but I felt that I was getting better at explaining things, and more insightful overall, as I taught, even when confronted by a young man who preferred swinging on the classroom's Venetian blinds to verse. I was a poetry fan, the way other people that year were Pearl Jam or Dr. Dre fans, and as such I believed that poetry had curative properties.

When you're young, you are not supposed to know who you are. Yet I knew who I was at twenty-three better than I do now. I would give up poetry, of course, because I knew even then that I would have to do that in order to stay in my version of the middle class. I knew I would have to give it up in order to make a living.

The reality for me, and for all the rest of America's contingent brain-workers, is that the security and respect once accorded to the lifetime academic—the promise of a middle-class life—just isn't there anymore for most. They no longer have a stable self-conception within reach. And an unstable class identity can be a great source of unhappiness.

Class mobility has long been seen as shorthand for progress; moving upward. People once trusted such mobility to get them out of their hometowns. They assumed that with the right training, they would be able to move into new, more comfortable, and more powerful existences.

But now this mobility works in the wrong direction for

some, who are sliding backward. After all, for so many intellectual workers, their refined training brought them huge school debt, not a better future.

Carla Bellamy, a professor of anthropology at Baruch College in Manhattan, agrees that this is the case. When I met her three years ago, she held a Ph.D. from Columbia University, earned $74,000 a year, and lived with her husband. She was pregnant and had a four-year-old daughter. Her husband was a part-time composer and the executive director of a music organization.

Bellamy's life took a certain difficult shape after her second baby was born. She used paid parental leave to care for both children. The family then went deeper into debt paying for preschool and day care for both children. Her husband didn't have a stable job. Bellamy had one foot in and one foot out of the middle class.

The lives of people like Bellamy raise the question: what does "middle-class" actually mean today? She appeared bourgeois, having gotten into a class, through education, that tends to have know-how and access; she had cultural capital, or nonmonetary capital, which is sometimes quite different from the economic capital of money and assets.

Bellamy had education and a mode of speech and dress that would seem likely to promote social mobility. She was a tenured professor who practiced ashtanga yoga, and she was far more trained and intellectual than her parents, who were part of an evangelical community in upstate New York.

Yet her parents had been economically better off than Bellamy and her family were now. They had owned their own home, for instance, even though her father was a bus driver. In the overheated real estate market of New York City, Bellamy and her family were likely to never be able to buy property. Bellamy would start to talk to other mothers like herself in New York City about her situation and realize that discussing your bank balance or even your weekly budget just was "not done." An inner voice would say, *Sorry to bring it up.* A delicate deerlike blonde, she'd go to the playground or the birthday parties and be suffused with self-consciousness. *How much of this is me needing to get over my economic situation,* she'd think to herself, *and how much is being genuinely disturbed by my growing debt?* She thought she'd lose friendships over such differences if she let them show. So she'd grit her teeth and say nothing about the "La La Land" of upper-class New Yorkers, with their gifted-and-talented test tutors for three- and four-year-olds. It depressed her, she said. How would she and her husband pay off six years of student loan debt on top of everything else?

In 2015, only an estimated 16 percent of women in the workforce, across all professions and the whole of the country, made $75,000 or more, so Professor Bellamy was privileged. (The percent of black women who make $75,000 or above is even lower today than the percent of white women: the former earn roughly 65 cents to white men's dollar, according to the Economic Policy Institute, which is roughly 16 cents less than white women, who make 81 cents to the dollar.)

But Bellamy didn't feel that way. Indeed, even with a combined household income of $110,000, she and her husband struggled to afford day care. It was a story I heard echoed when I spoke with other female professors, who sometimes took sick days even when they were healthy so they would not have to pay for babysitters.

"Our entire disposable income goes to child care," Bellamy said. "It's not a tragic story, but it is tiring and tiresome. I have a career, I work really hard, and yet I get no break."

Bellamy had considered working at a restaurant as a hostess during summer vacation, but since this is the time academics typically use to do the writing and research that ultimately help them get tenure, she wound up not doing so. But spending her summers in academic pursuits put her family at greater financial peril. (I spoke to another adjunct professor parent in the suburbs who worked as a waitress at an Italian family restaurant on the side and was embarrassed when she encountered her students while serving them. Her story reminded me how much class anxiety is tied up in the fear of humiliation.) Bellamy compared her social life to a Jane Austen novel. The women she knew had made it into or out of the urban middle class thanks to their choice in husbands: the wealthier the partner, the more they flourished, and the less they seemed to understand the plights of others.

In the meantime, she had no hope of a raise, as a wage freeze was in place at CUNY, the New York City public college system that employed her, due to budgetary issues. "I'd go out to eat every six years!" she said.

In time, her husband finally got an additional better-paying job as an organist in a church for $50,000 a year. This brought their combined income up to $130,000 or so, which was good news, of course, and more recently it had risen again to around $160,000 when she became chair of her department and finally got a raise. But their combined earnings now also blocked their access to middle-class affordable housing under New York City's Housing Development Fund Corporation (HDFC) program, which would have enabled them to move. Their ineligibility for affordable housing brought on a whole new wave of anxiety about money. Bellamy thought about it every day. She had a visceral disgust for the anxious competitiveness of other middle-class mothers on the playgrounds and in the preschools. "Who cares what skills my kids have mastered?" she told me.

Over the years I watched as this modest oppression politicized Bellamy. Now forty-five and the mother of an eight-year-old and a four-year-old, she had become an intense Bernie Sanders supporter during the 2016 election, going to his rallies in the Bronx, scrambling to pay for day care for her hours working for the campaign, phone banking for the candidate, going door to door. "He's been saying things I've been feeling for so long that no one else will really say," she explained as she pushed her daughter in a swing on a Harlem playground. "I've been so isolated with my feelings and thoughts about inequality that I've become obsessed with him." Would she be able to pass on the social class standing she had created for herself? Would they even be able to get the same fine education she'd gotten as a bright and dedicated young woman, vaulting into the Ivy League after growing up deeply religious and attending an equally devout college?

Still a Bernie Sanders supporter a year after the election, Bellamy bristled at the memory of the 2016 primaries. She was now researching her new book project, "Jyeshtha, the Hindu God of Misfortune." Misfortune was indeed a subject that for her would seem apropos for our times.

THERE ARE BOTH SMALL AND LARGE REMEDIES FOR THE PLIGHT OF the hyper-educated working poor—those earning around $36,000 a year, with kids, and just getting by, only a few false moves away from the poverty line. For underpaid and often desperate adjunct instructors, one particular remedy is what I think of as "unusual unions." The last five years have seen a rise of atypical union members, like adjuncts, despite the downturn in labor membership overall. And unions have started organizing precarious workers, from adjunct math professors to fast-food servers.

Take the former: today more than 40 percent of the teachers at American colleges and universities are adjuncts. They are typically paid by the class, like Bri Bolin was. To try to get some protection in these depressing circumstances, adjunct faculty members at Duke University recently joined the Service Employees International Union (SEIU). In 2013, adjuncts at Tufts and other schools decided to unionize. With a national median salary of $2,700 per class without benefits in 2010, it's no surprise that the professors I spoke to were like Mary-Grace Gainer, an adjunct professor who spent half her family's household income on rent and couldn't afford to have an extra visitor over to dinner. (Gainer's husband intermittently drove a school bus to make ends meet.) In the family's part of Pennsylvania, fracking had also jacked up the rents, so after her tight finances got even tighter, the family had to move to Lock Haven, Pennsylvania—"the Boondocks," as Gainer put it, where a rental still cost only $2,000 a month. Meanwhile, Gainer drove roughly sixty miles on teaching days to get to each of the several different colleges she taught at, spending her rare dollars on gas. She earned a total of $36,000 yearly. One year Gainer had to save for weeks to buy her five-year-old a birthday cake.

And this is why a union representing some adjuncts, the SEIU, has set a pay target of $15,000 per course, in total comp and benefits. Uncommon unions could put a dent in the present university business model and eventually free some brain-workers from desperation and penury.

Another quirkier correction is what I call "the fair labor seal." This commendation would be given to, say, colleges that treat their employees well. The labor practices of colleges and universities would be rated for their fairness and for how precarious their workers are. This strategy would work because American parents remain spellbound by *U.S. News & World Report*'s ranking of colleges when they are searching for schools for their children. These ratings tend to take into account factors like the SAT scores of those admitted and the number of those who apply versus the quantity the colleges accept. But what if the percentage of adjuncts showed up in such a ranking of universities and was factored into the overall quality of the education? What if the amount adjuncts were paid and how many classes they taught were also factored into the rankings? How would these rankings change if the best colleges and universities for employing full-time or tenured faculty or, at the very least, decently paid adjuncts with some degree of job security received the fair labor stamp of approval (or disapproval)?

If this "seal" was adopted, colleges could then call themselves "fair labor" institutions the same way that some brands call themselves "fair trade." Since many universities have claimed that much of their money goes into attracting students in a competitive marketplace (and their parents), such a stamp would give them a certain advantage and also incentivize better treatment of their underemployed faculty. A hurdle to overcome, of course, would be the unwillingness of some university administrations to share this information. What if the adjunct faculty themselves at institutions adjudicated whether their colleges could receive the labor stamp?

A third solution is smaller-scale—what if alumni donors organized to earmark their gifts to improve the lot of adjunct faculty? Where alumni could do this, what if universities were obliged to pay attention to these conditional donations, especially smaller donations?

My favorite—and perhaps the most bespoke—solution to

being hyper-educated poor is one that Bri Bolin came up with herself.

When Bolin's friend Joe Fruscione was an adjunct professor at Georgetown University, he had to occasionally sell his own furniture and video games on the website Half.com for extra cash. Bolin had taught at Columbia College in Chicago for many years, yet she often had to buy her groceries with food stamps. Fruscione and Bolin were not alone, of course. The situation is so bad that Bolin, along with Fruscione and another teacher—Kat Jacobsen, who goes by the *nom de adjunct* Kat Skills—decided to create a nonprofit devoted to helping impoverished professors. To conjure their financially precarious situation and that of their comrades, they called it PrecariCorps.

First they set up a website and announced its mission: "Agents for Higher Ed: Seeking to provide temporary, welcome relief from the economic, emotional, and physiological stressors that all too often define the life of an adjunct educator." Through the site, anyone can make a donation. They are creating an application process that will weigh the merits of requests for aid for research materials, or travel for said research, or even for medical bills. As of 2017, the trio had received over one hundred donations and fifty requests for funding. (They told me that they draw no salaries for themselves.) They had also distributed funds—$10,000 so far—to select adjunct professors living on the economic edges.

"I was scrambling, tutoring, and nervous," Fruscione said of his life before he met and married a woman who had a well-paying job. "I was lower-middle-class at best, with little savings, IRA, or retirement. I couldn't have adopted a child on an adjunct salary, as my wife and I are now doing. I wouldn't have passed the requirements." (They finalized their child's adoption in 2016.)

It got so bad for Fruscione that he finally quit academia and changed professions. He now worked as a copyeditor in the vaguely defined world of writing consultants, even though he had a Ph.D. from the highly regarded George Washington University and taught there—and at Georgetown—for many years.

Bolin, who was clearly still making her way with difficulty, had a little bit of a break. A handful of strangers sent checks and anonymous gift cards to her—via email, her faculty mailbox, and her home address—including a single $5,000 donation, after I published a piece about the financial straits she was in. Someone even donated a tricycle to her disabled son, Finn. Bolin shared some of the money, she told me, with two needier adjunct friends she met through online activism. They both were also mothers.

It had been all of this generosity that convinced Bolin to create something a little more formal in PrecariCorps. Her colleague wanted to be able to pay her electric bills or go to a conference or buy a textbook—and they both thought other adjuncts should be able to do so too. They had started having conversations about creating some kind of group like PrecariCorps with Fruscione, whom they met in 2014 via Facebook academic activist groups. They also created PrecariCorps to draw attention to their profession's plight. Adjuncts are, after all, veritable poster children for the erosion of the middle class. These days, professors may be more likely than their students to be living in basement apartments and subsisting on ramen and Tabasco.

There is no shortage of professors out there for PrecariCorps to help, including the adjunct mom I mentioned earlier who waitressed at a local family restaurant on the weekends. Or the medievalist on Medicaid. Or Mary-Faith Cerasoli, the "homeless professor" who lived in her car. (I got in touch with Cerasoli a few years ago after she went on a hunger strike to draw attention to faculty poverty. She halted the strike after six days.) PrecariCorps also received a request for funds from an adjunct whose university took so long to pay him that he started accumulating financial penalties from his bank for overdrafts.

PrecariCorps is scrappy and fledgling, like a DIY benevolent association. But it is also part of a larger movement to defend adjuncts—organized by the SEIU, which includes hospital workers and janitors—that includes sizable groups such as the Coalition of Contingent Academic Labor, the New Faculty Majority, and a group called Adjunct Action. In February 2015, during National Adjunct Walkout Day, thousands of adjuncts, general faculty, and students walked out of their classes on both coasts in a plea for fair wages and better working conditions. After all, adjuncts tend to do the same teaching work as tenured professors and usually have the same credentials.

When confronted with the adjuncts' lot, university administrators often claim that they are in a bind. They point to budgetary realities—the shortfalls that necessitate having only adjunct positions if tuition increases are to be avoided. They note that the public has demanded greater accountability: America's students and their parents are angry that education costs have gone up exponentially.

But why has the cost of education risen so fast? In 2013, a raft of articles and studies found that tuition at colleges and universities was increasing faster than inflation and pinned the blame on the fact that public universities had been hiring far more administrators than teachers, creating sprawling bureaucracies. In fact, according to U.S. Department of Education data, college and university administrative positions grew by 60 percent between 1993 and 2009—ten times the rate of growth of tenured faculty positions. Why weren't universities using their pricey tuitions to pay the adjuncts, whom students actually saw each day, rather than the costly administrators?

In an attempt to right this skewed situation for contingent teachers, the adjuncts' rights movement has been pushing for state legislatures to impose binding contracts on state colleges and universities to provide health and retirement benefits to adjunct faculty who work part-time hours or more. The organizations in this movement have made some inroads: in Colorado, for instance, a bill seeking to end the "two-tier faculty system in Colorado's community colleges" circulated around the Colorado state legislature. That bill would have made Colorado the first state to render illegal what is now sometimes awkwardly called "adjunctification," but it was defeated in 2015.

PrecariCorps offered a different strategy. As Bolin, Fruscione, and Jacobsen tried to get serious about fund-raising, some PrecariCorps members attended conferences with their proverbial wool hats in their hands, beseeching the most comfortable of the tenure-track faculty for donations. The logic: established, financially stable members of a once-middle-class profession could partly subsidize their destitute sisters and brothers. Ultimately the group hoped that colleges would be forced to change their deficient labor practices. But in the meantime, they coaxed the academic system's elite—the tenured—into solidarity, hoping they would open their Italian leather wallets to support the worker-bee adjuncts who propped up their comfortable world. Would it work? Only time—which these instructors were paid so little for—would tell.

BOLIN AND HER FRIENDS AND COLLEAGUES SHOW US THAT THE middle class as a symbolic or existential category is under siege. In less than a generation, transformations have remade their trajectory. Corporations have grown exponentially and globalized; these giant companies, though legally considered people, now may dehumanize actual human beings through computer scheduling systems that limit hours, treating employees as mere contract workers with no benefits. (You will read more about this in the next chapter.) This practice extends to for-profit colleges, which, of course, are corporations. With wage stagnation, incomes no longer support a family as they might have in the past. But this labor situation raises a more psychological, even existential, question, one that returns us to the nearly impossible edict to "do what you love." "What happens to optimism when futurity splinters as a prop for getting through life?" asks Lauren Berlant in *Cruel Optimism*. She could well be asking this question of the adjuncts and the other Middle Precariat brain-workers you have met and will meet in *Squeezed*: Who are we? And what are we looking forward to without linear or even curvilinear professional futures? In the past, we might have been ambivalent about security in the workplace—seeing it as the "alienated prison" that the theorist Max Weber described—but what about now, when we are supposedly full of affection for what we do for work but so precariously employed that we are actually quite detached from it? Has the advice to do what you love become a curse to some? And if so, is that unacceptable?

The young, upwardly mobile professional parent can now seem like a type of the past, replaced by a generation whose income has stalled. For the middle class, the largest group of working people in the United States, stagnation is experienced as a great loss.

Finely educated minds feel themselves at a loss in a new job market. They are different from the graduate students and adjuncts who lived on lentils in the past—or only on iced coffee, as I did twenty years ago teaching at a community college. Back then, there was still fading hope that a tenure-track job would ultimately appear on the horizon. That this is no longer a realistic hope is disturbing. Historically we have believed that we need an intelligentsia if we are to have a middle class in the first place. In fact, the intelligentsia have sometimes been thought to be the "spokespeople" for the bourgeoisie, for better or worse.

When I checked in with Bolin again, she had stopped teaching at the college, after eleven years or so, and in 2016 she was starting a new grad school degree.

Finn was walking a little more now, his halting steps greeted with much jubilation. Bolin had started training as a speech-language pathologist focusing on alternative communication. She got depressed the last time she tried to pursue this profession, but now she was excited again at the prospect of working with nonverbal people like her son, teaching them to communicate through technology. Bolin was trying to come up with $70,000 in tuition. "I'll be earning a degree that will support myself and my son financially and be meaningful and ethically sound. It's been the most difficult therapist to find for Finn, so I'm here to help fill that gap."

About a year later, in 2017, Bolin was enrolled in a full-time speech pathology program. She now qualified year-round for food stamps and still received disability payments for Finn. She also now got $230 per month from Finn's father. She had paid off her credit card debt by borrowing from family and having her loans for school deferred. What remained was a credit card balance of $2,000 or so. Finn, now eleven, was in a school for children with disabilities, paid for by his public school, and getting therapies, except for speech.

Bolin recently raised money on GoFundMe to pay for horseback riding lessons for Finn. She was even off Xanax for her anxiety, she said, after she left the adjunct professor racket behind.

She gave notice at the college where she had taught for so many years and was told, "We're sorry to see you go"—but little else.

3

EXTREME DAY CARE

THE DEEP COST OF AMERICAN WORK

In the garden of Dee's Tots Child Care, amid the sunflowers, cornstalks, and plastic cars, a three-year-old girl with beads in her braids and a blond two-year-old boy were shimmying. These were Deloris Hogan's 6:45 P.M. pickups. Nearby, also dancing, were four kids who wouldn't be picked up until late at night, as well as two "overnight babies," as Deloris called them. Dee's Tots stayed open twenty-four hours a day, seven days a week, and this irregular cycle of drop-offs and pick-ups resulted from the unconventional hours worked by the children's parents.

One afternoon in August 2014 the kids bounced on the center's inflatable castles, rustled around at the sand tables, and ate a watermelon snack. Then it got dark. By 8:30 P.M., three-year-old Naima was in her pink polka-dot pajamas. Little Ivette and her watchful older sister, Diana, were lying on thin mattresses laid over yoga mats. Even with the lights dimmed, you could make out the bright posters on the wall, the costumes and the boxes of dolls and baby clothes, and the fluttering rainbow curtain dividing one room from the other. The Genie was riffing as *Aladdin* played in the background.

Deloris changed two-year-old Kaden's diaper; next, she gave one-year-old Noah a bath. Watching the bedtime endgame of "extreme day care"—my term for the rapid expansion of day cares with evening or early-morning hours, or even all-night care—it seemed to me to be a fretful process. I feared that even efficient, poised Deloris wouldn't be able to wash all the kids in time for bed, and that I'd have to help herd her wards to the bathroom myself, and then to their pallets. What if the children turned into tearstained goblins of exhaustion before Deloris got them tucked in? How would she manage it? The main room at Dee's Tots looked like a supersized slumber party, but the truth is that this was an ordinary day.

Dee's is only one of a number of twenty-four-hour child-care centers around the country. Just on this single block in New Rochelle, New York, there was another facility, Little Blessings, that offered overnight care as well. During the week I spent there, Little Blessings and Dee's were in a nearly comic "decoration-off," competing for kids with colored lights and giant Doras and Spider-Mans. By contrast, some round-the-clock centers make their pitch more toward parents, with aspirational names like Success Kidz 24 Hour Enrichment Center, in Columbus, Ohio or Tip Top Child Development Center or Five Star Sitters, both in Las Vegas.

The growth of this industry reflects the upheavals in American work lives. A huge and growing proportion of us now work an expanded workweek with unpredictable working hours. This reflects, among other things, the 24/7 business environment of the twenty-first century: both the digital economy and the freelance and gig economy, as well as the failure of wages to rise to match the growth in the economy, meaning that wages haven't kept up with costs, even at a time of low inflation. The rise of 24/7 day care also reflects the disempowerment of unions and with that the extreme corporate schedules that have shattered the traditional workweek and hours for employees. And to call it "disempowerment" is not an exaggeration. Although 30 percent of Americans were in unions in the 1960s, now only 11 percent of all workers are union members, and only 7 percent of private-sector employees are union members. Unions have lost bargaining strength and also popularity among American voters.

As of 2004, nearly 40 percent of Americans had experienced nonstandard work lives, if by "standard" is meant the (now semi-mythical) eight-hour daily shift of the past. According to the National Women's Law Center, 9 percent of day-care-center care is now provided during evenings or weekends, more than enough to compose a trend story. In addition, almost two-thirds (64.2 percent) of women with children under age six are working and one in five working moms of small children work at low-wage jobs that typically pay $10.50 per hour. They all need to earn more if they are to truly afford day care.

The average American adult also now works forty-seven hours a week. Working people who live below the poverty line are particularly afraid to say no to these unusual schedules. They may have no one to say no to, anyway, as some employers use data and algorithms rather than human managers to create schedules in the hopes of saving money, supposedly so that workers will spend fewer hours "sitting around." The software doesn't care if a shift falls in the middle of the night, tearing a big hole in employees' family lives by preventing them from getting home to put their kids to sleep, or to make them breakfast, because they've been scheduled to work the earliest shift.

The extreme day cares of today reveal how treacherous and downright absurd work schedules can be in this country. We now need twenty-four-hour day care. Indeed, extreme day care represents the ways in which the hours we are asked to work now squeeze us.

Well-paid professionals who work evenings may be able to afford one or two nannies, or they may have nonworking partners who stay at home. But parents like the ones who rely on Dee's can't afford such luxuries. Diana and Ivette's mother, Marisol, for instance, was raising the girls on her own, working at a supermarket from 8:00 A.M. to 2:00 P.M. and at Home Depot from 6:00 to 10:00 P.M., six days a week. The girls were at Dee's Tots for both of her shifts, and she was with them between 2:30 and 5:30 P.M. each day. "I worked one job twenty-nine hours a week, so I got a second job, as I can't afford to take care of my kids—I need more money to be surviving," said Marisol, a slim young woman with glasses and pulled-back hair who came to the United States from Mexico when she was four. Marisol works twenty-nine hours at each of her jobs. This is common. If an employee works over thirty hours, the employer is required to provide health insurance. (Not surprisingly, the number of part-timers working just below thirty hours a week rose from 2013 to 2015, and the number working just over thirty hours fell.)

"With car payments coming up, I applied for Home Depot," she said. As a single mom, she needed to have some serious backup. "The Hogans are very open about my schedule and were willing to work with me."

Marisol missed a lot of her kids' growing up—a particularly difficult reality since she feels she has to do these extra jobs for them. One afternoon I saw her at Dee's Tots carrying in a tray of chocolate-frosted cupcakes with pink decorations for Ivette to celebrate her fourth birthday with her day- and night-care friends. "I have to wake them up. With the little one it's easy, but the older one . . ." Marisol trailed off. Deloris and her husband, Patrick Hogan, who co-own the center, had nicknames for all the kids—Chowder, K.K., Little Bit, Jelly. Most spent more time with Patrick and Deloris, whom the kids call Nunu, than with their parents.

Many of the kids at the day care were African American, which made some statistical sense: African American women have worked at higher rates than other women for decades. Child care is very much needed by these families.

The Hogans started this day care in 1985 to earn a living while looking after their own children. Now Deloris and Patrick, ages fifty-seven and sixty-two, respectively, work around the clock. Those who provide child-care services for the folks with crazy schedules have crazy schedules themselves. Both go to bed around 1:00 or 2:00 A.M. Deloris then wakes up again if one of the overnight babies needs attention. She goes back to sleep and Patrick wakes up at 5:45 A.M. He makes breakfast for the overnight babies and prepares the house for the 6:00 A.M. drop-offs. (Nobody comes or goes at Dee's Tots between 3:00 and 6:00 A.M.)

Deloris and Patrick met in 1974, in the housing projects across town from the home they now own. Patrick's mother raised him; his father, a prizefighter and a fry cook, was not around much. Both Deloris's parents brought her up in rural Mississippi. They picked cotton and tobacco on former plantations. Deloris and Patrick had been together for forty years. They were proud of their four children, some of whom had already graduated from college and embarked on professional lives. They were also proud of Dee's. Patrick wore hospital scrubs in many different colors—red, blue, purple, yellow, and tan—as a sort of professional uniform. "It's good for branding," he said. (To his delight, he was also sometimes mistaken for a doctor.)

The Hogans are African American, in a sense emblematic of the child-care workforce, which is disproportionately composed of people of color. But Patrick is an anomaly in another way: as of 2015, 95.6 percent of those doing the caregiving were female.

Occasionally, Patrick's pride in his work had an edge, as when he'd serve the kids food and joke with them—"Six dollars, please"—as if he were a waiter and these children were his customers. Which in a sense they were. I asked him if he ever tired of his extreme work schedule.

"Would I enjoy not working? Yes," he said. "But who wouldn't?"

Six months in advance, the Hogans would prepare for parents a list of days when Dee's Tots would be closed. They closed the day care for fifteen hours each year for job training, as required by the state, and each year they went away for their anniversary, but they didn't always take off for Thanksgiving.

"The stores and the nurses are not off for Thanksgiving," Deloris said. Still, her caretaking commitment, so ample that she offered me tips for getting my own daughter off bottles, had its limits. One child's mother was asked to work a shift starting at 3:30 A.M. The mother asked the Hogans if she could drop off her child around that time; the answer was no.

The Hogans understood that they were in a supplemental relationship with the children's parents, and that the children's well-being, as well as their own financial security, depended on their providing around-the-clock nurturance. The demand for this kind of extreme day care will only grow as the number of nursing jobs, for instance, rises. According to U.S. Bureau of Labor Statistics (BLS) employment projections, registered nursing as an occupation will grow by 16 percent through 2024, owing to a large and growing senior population. (Of course, as this is squeezed America, all is not well: we will read more about the problems that arise for nurses later in this book.)

Hospitals and other conglomerates, like Target, force irregular hours onto their lowest-paid workers, sometimes with no sensitivity to workers' needs. Consumers too play a role. We assume that we should be able to buy or eat whatever we want, whenever we want, even in the middle of the night. Extreme day care has risen up in part because our system does not ensure that the needs of all families are met, including those parents who work odd hours.

Ours is a forever clock.

TWENTY-FOUR-HOUR DAY CARE CAN SEEM SHOCKING TO PEOPLE who have no need for it. When I mentioned extreme day care to some parents in my own child's day care, which ended at 5:45 in the evening, a number recoiled. These parents had their own children enrolled full-time at tonier day-care centers or employed a nanny. They also often worked far into the night, laptops aglow, making their dimly lit homes look like aquariums. Yet many found it strange to have a child at a facility overnight. A number were surprised to learn that such places even exist.

But of course they do exist. The relationship between employers and employees has undergone a gloomy revolution in this millennium, much of it to the employees' detriment. Employers often may not see or accept an obligation to make workers' lives bearable. The term "standard of living," once all the rage among American sociologists and historians, is used less and less. (Perhaps the notion that a relatively high quality of life should include small pleasures and comforts has faded.)

This is not to say that the 24/7 job or evening work was invented in this decade, as an army of nurses and night watchmen, waiters, and doctors can attest. The more privileged may have au pairs and sitters to pick up the slack, and perhaps day-care concierges or assistants to simplify their search for care in the first place. ("Life would be so much easier if we had personal assistants for our children," exclaims the website for one such company, Kid Care Concierge, of Bridgewater, New Jersey. Kid Care Concierge, a "personalized and exclusive concierge service firm, establish[es] a harmonious work-life balance for busy families.") For the more affluent, there's now even an on-demand ride-and-care service for children in Southern California and the Bay Area called Zūm; an expensive car service like Lyft, Zūm ferries children ages five to eighteen to and from school or to team practices and music lessons, with child care provided before or after the rides for an additional fee. But such services are not affordable for late-night store and chain-restaurant workers. Indeed, as Rachel Cusk writes in her memoir of how the wealthy deploy day care: "The deep-pile nanny, the nanny who exists to cushion the impact of parenthood, was, I discovered, the preserve of the wealthy. All other forms of childcare appeared to operate on the principles of a public callbox."

The culture of erratic and endless work afflicts so many of us—both faltering middle-class parents and the professional caregivers who make it all possible as they work in homes or day cares for little more than minimum wage from morning to night. These parents and caregivers coalesce like nesting dolls in a symbiotic relationship as they try to adjust to the horrifying, fresh element here: just-in-time work.

"Just-in-time" scheduling sets peculiar and erratic working hours that serve the company's needs with little regard for those of the employee, such as a sudden call to an employee to work a shift, with little notice—now an ever-present possibility in our culture of overwork. Besides being called into work on short notice, workers may also be expected to work hours that fall outside the typical nine-to-five schedule. Indeed, in 2015 unpredictable work schedules governed the lives of 17 percent of American workers, according to the Economic Policy Institute. Some companies, such as J.Crew, Urban Outfitters, and the Gap, have responded to intense pressure from media and advocates by ending just-in-time schedules. Despite a federal bill reintroduced after an earlier failure in 2017 and attempts at legislation in ten states to require employers to stabilize their workers' schedules, erratically scheduled jobs remain widespread.

As the labor specialist and CUNY professor of history Joshua Freeman told me, "Only a tiny number of Americans now have a normal five-day, forty-hour workweek." A 2014 Gallup poll had similarly disturbing findings: although "full-time" is still defined as a forty-hour workweek, not even four in ten Americans employed full-time work this little. In fact, 42 percent work forty hours, and roughly 50 percent work more than forty hours: the average is forty-seven hours; only 8 percent of full-time employees say that they work fewer than forty hours. (Eighteen percent work sixty hours a week or more!)

Some corporations now expect worker hours to conform to round-the-clock production schedules that have been tweaked for maximum efficiency and to accommodate the time zones of business partners halfway around the world (how do you say "overtime" in Hindi?) and real-time fluctuations in demand and numbers of customers. Companies ask employees to be available at the last minute exactly when needed, with little regard to how scheduling affects their family lives. Parents who work such hours often aren't home to help their children get their science projects done or look over their math homework. Living by the forever clock, families subject to just-in-time schedules never have enough time. One important reason American families are used by such schedules is that our leaders, though they say that they value parenting and caregiving, do little to support them and much to impede their efforts.

The Hogans work hard for little income at all hours of the night, and the parental penalty punishes parents of limited resources by not supporting them as they drive back and forth between their different jobs and the day-care center in a carousel of frantic activity. The caregivers and parents—and their struggles—are lodged within each other, like nesting dolls.

Extreme day care and the crazy, brutal work demands that make it necessary lay bare the extraordinary difficulty of arranging reliable care for our children. Over the last three years, I talked to hundreds of parents who felt like they ought to have been able to afford child care or provide it themselves. Weren't they competent administrators of their own lives? Still, they constantly found themselves coming up short, and it was often child care and its costs that nearly did them in.

Foremost is the shock of how much day care costs. If middle-class parents have become an endangered species, pressed in on every side by harsh work-family policies, the cost of child care is very much to blame: the United States is close to the bottom of the ranks of wealthy nations with respect to our federal child-care expenditures as a percentage of gross national product (GNP). Joya Misra, a professor of sociology at the University of Massachusetts Amherst, has analyzed data from thousands of parents from different social classes. One study of middle-class academic parents was based on hundreds of surveys and focus group interviews and seventeen one-on-one interviews. Many survey participants talked about the shock of day-care costs, which can devour 30 percent of one income in a two-salary household, Misra said. In thirty-three states and the District of Columbia, the cost of center-based day care (let alone a nanny) for an infant is higher than the cost of a year at a public college. I'm told that day-care costs for middle-class New Yorkers can easily add up to $25,000 to $30,000 per child. In New York, child care is the single greatest expense among low-income families, surpassing both food and housing costs.

Parents may be so desperate for ways to pay for their children's day care that they advertise on fund-raising sites: "A single mother and full-time professor, caught in a pinch between salaried work and when coverage starts. . . . After a series of unexpected emergency expenses, I'm unable to come up with the necessary money in time." That mother eventually raised $2,286 of her "$2,500 goal" and posted a sweet picture of Sebastian, thirteen months old, at his interim day care wearing a bib. I couldn't help pondering how downright bizarre it was that Sebastian's mother was driven to ask for donations to access a basic service.

Beyond the stress of paying for day care are the inherent stresses, on both workers and their families, of extreme schedules. For instance, when one mom I spoke with got to work at a McDonald's in San Jose at 7:30 A.M., when much of the city was just waking up, she would have already dropped off her youngest son at day care. By 7:45 A.M., she was flipping eggs for McMuffins and cleaning off tables. When her workday ended more than eight hours later, she still couldn't rest for even a moment: at 4:00 P.M., she would rush to pick up her four-year-old from day care and her twelve-year-old from school.

If this mother was late picking up her little one, the day-care center was supposed to charge her a dollar a minute. (She said that they had recently warned her that if she was late again, they would actually do so.) If she was slow to pick up her older child, who waited for her in his school's library, his homework often suffered. While he awaited her pickup, there was no one there to help him and ensure that he completed his assignments, which would get squeezed later between a late dinner and the range of bedtimes his mother has to supervise. A single parent, she was often late in picking up her children because her manager would ask her to do extra tasks at the end of her shift, keeping her from leaving on time. These changes would add to her stresses: she worked around thirty-five hours a week, but those hours still didn't cover her expenses, and she received no child support from her children's disengaged father. The family lived in a garage, and the tight quarters were hard for her growing boys. Her third child, her fifteen-year-old son, wanted his own room, away from his two little brothers. And her lateness had affected her middle son's grades, which had started to plummet.

Her son needed materials for a project involving the solar system: paints and Styrofoam balls to represent planets. But her boss changed her schedule, and she came home late from work once again. Because of her unstable hours and inability to afford adequate day care, she was often not around to help him with his homework, and as he missed assignments, his usual As and Bs had been dropping.

The other party who suffers from the lack of state support for day cares and bears the brunt of the new twenty-four-hour labor regime is the day-care worker. We continue to overlook these workers' needs. Part of the problem is our culture's resistance to the idea of "day care" itself. There is a widely held notion that, because America doesn't support a public, nationalized system of care, day care should be wholly funded by the private sector. (In this sense, day care resembles the health-care system, a private-sector system that fails because of social resistance to its institutionalization.) As Kathleen Gerson, a sociologist at New York University, put it when we spoke: "What is our strange commitment to seeing care as a purely private individual act rather than one embedded in the larger community?"

Day care itself is also specifically stigmatized, a holdover from the conservative backlash against working women of the 1980s, when day care became a Satanic straw man for so many societal evils. As Richard Beck argues in *We Believe the Children: A Moral Panic in the 1980s,* the sexual abuse trials against day-care center workers in that decade stemmed from a reactionary fear of feminism and of women working outside the home—with the concomitant need for day care—as well as the usual fear of crime. The scapegoating of day care also stemmed from societal dismay over what many perceived as the dissolution of the traditional family, with mothers waiting at the school gates for pickup and fathers home by six, taking their seat at the head of the table.

Caregiving also is the object of a more realistic critique, as some have noted the psychological toll of the profession. Scholar Arlie Hochschild, who writes about day-care operators and workers like the Hogans, worries about the potential harm to workers who must sell the most intimate parts of themselves, manufacturing smiles and cuddles for low pay. Her book *The Outsourced Self* shows the history of this: that what was once part of private life, from love to child-rearing, is now outsourced.

But there is an even broader reason that care and care work are looked down upon. The theorist Jeremy Rifkin puts it well when he writes that our degradation of care is caused by "hypercapitalism," our crazily unfettered free market. Rifkin channels my own worries and beliefs when he argues that we substitute market transactions for what should be human interactions. "But when most relationships become commercial relationships . . . what is left for relationships of a noncommercial nature?" asks Rifkin. In this transactional American life, human relations are only "held together by contracts and financial instruments," and reciprocal relationships "born of affection, love, and devotion" are kaput. This is a part of what has been called the "commodification of emotion"; when we sell our most intimate relationships, the result can be emotionally (and monetarily) damaging. On one hand, we can start to mix up love and money. On the other, or we can wind up severely underpaid, like the nannies or nurses or day-care workers who care for others for a living who are subject to what is sometimes called the "prisoner of love phenomenon."

The "prisoner of love" theory asserts that employers can underpay care workers in part because they know they can manipulate these workers' specifically caring natures. As the market economy expands still further, now occupying all corners of our lives, the care of children as well as the elderly is becoming ever more degraded. The basic reason for that degradation, beyond the mere transactional nature of day care, is clearly the sexist view of caregiving as "women's work" and thus as less serious, less valued work that in the past would have been handled for "free" by the traditional stay-at-home mother.

To my mind, the contempt for care work has even broader dimensions, with roots in everything from religion to American notions of what comprises achievement. The historic origins of this contempt: care work has long been associated with charity and piety. Care workers were either not paid or paid in alms. Regular salaries would defeat the long-standing spiritual valence of care work as self-sacrifice for others.

As for the unworthiness of care work, it's as if we Americans assume that when people spend their time loving or tending to others they have somehow stepped out of the commercial—and thus any pragmatic, valuable, and intelligent—sphere. This is partly why these workers are paid little; indeed, they are made to *pay* for what is seen as their unworldliness or idealism. Low pay and lack of respect are, in a sense, what they earn for having jobs taking care of America's most vulnerable.

But what if we could combine care and money? This "love and money" framework argues that markets are not the opposite of true care. Viviana Zelizer, a Princeton University scholar, thinks that we first must understand that no *contamination* results when the two spheres of intimacy and the marketplace overlap. The ideal, she writes, is to blend "intimacy and economic activity . . . in constructing and negotiating 'Connected Lives.'" On one side, parents seeking day care for their children may separate nurturance from success, worrying that considerations of economics and money could sully and degrade love, especially for children. On the other side, they may arrive emotionally armored in the transactional workplace, believing it is best if that space is as free of sentiment as possible. But what if we could bring the two sides together? If that happened, Americans might pay their care workers better.

My own experience with my daughter informs my view of the love-and-money division. Like many parents, I couldn't go back to work until I had employed caregivers I trusted.

I went on sitter sites—meeting women on these sites was akin to dating apps. And one time it clicked: I found a young woman on the site who wasn't smiling brightly into the camera and looked thoughtful, her make-up-free face framed by a blunt-cut bob. Her name was Sydney and she soon had my daughter dancing in her crib to obscure 1960s music emanating from my broken iPhone. I was paying for my freedom with money. But I also paid emotionally with the new distance from my daughter's body: we had become what sometimes felt like a single pale organism of cartilage and skin, and when I was separated from her, I wasn't quite sure what I was without my literal other half.

In the course of that year, my daughter fell deeply in love with Sydney, retro indie rock, and games of endless, absurdist peekaboo. In a rare collision of intimacy and economics, I could go back to work knowing that she was being cared for in a way that was often more energetic and freewheeling than anything I could give her. I could freelance-edit at a nonprofit office away from home, still pumping milk. It was liberating, if a bit alienated, being able to do this. After Sydney got a teaching certificate and stopped nannying, we decided to send our daughter to a day care with classrooms lined with orange and blue toddler paintings in drugstore frames. As for so many others, it was an effort making the monthly payments at that little Eden.

There was a history to our mini-plight. Today's day-care struggle for middle-class and working-class families was set in motion in the past. The first was in 1971, the year before I was born. The second occurred later that same decade, when work hours started to be contorted and extended beyond the standard nine-to-five, five-day workweek and our "needy workplaces" began to impose their expectations. It wasn't that children's and women's rights defenders didn't try to change these workplace practices. In fact, these advocates scored a watershed victory when Congress passed the Comprehensive Child Development Act of 1971, as part of the Economic Opportunity Amendments. This bill addressed the rising demand for day care as more women went to work. It also promised to improve the educational quality of day care, from early infancy to early adolescence, and would have subsidized care for working mothers and even for some of the middle class. But President Richard Nixon vetoed the bill later that year, and the dream of truly equal families was replaced by a chillier reality. Earlier, in 1970, the House of Representatives had voted in favor of a guaranteed annual income for all needy Americans, an offshoot of Nixon's Family Assistance Plan, because even back in the seventies, manufacturing jobs were already being computerized and people were grappling with the end of work as they had known it.

AFTER I HAD SPENT HOURS AT DEE'S TOTS, LONG INTO THE NIGHT and sometimes even into the next morning, I started looking for solutions for our twisted child-care complex.

The most obvious measure is to make child care affordable through subsidization. However, political resistance to this solution is overdetermined and long-standing. The prospect of advocating for it is wearying. One class of measures that appear to be next to useless would be the tax deductions and rebates that are standard fare in Republican tax plans. It's hard to see how such deductions, which amount to very little for scrambling working-class people, would help with essential services such as child care. Suppose a family pays $20,000 a year, or 30 percent of their income, for child care (not an unusual circumstance): how would a deduction help this family? For poorer people, such schemes are even more useless. One plan features a rebate of $1,200 a year for day care for lower-income families, which is a fraction of actual day-care expenses for most families.

The lack of adequate federal support for child care illustrates the devaluation of care work in America, which burdens both care workers and the middle-class workers who need to make ends meet by using the labor of underpaid day-care workers.

People in other countries across the world count on the availability of day care and see it as a collective good. In France, the government provides reasonably priced day care as well as tax breaks for families with in-home au pairs or nannies—plus universal preschool starting at age three. Some friends of mine left my city to move to France with their young children simply to have access to these policies and *crèches,* the French term for nurseries. Others moved to Germany because it offers generally more obtainable and better day care. Later, Germany passed a law guaranteeing every child over twelve months of age a slot at a day-care facility in the hope that the policy would help reverse its birth rate, one of the lowest in Europe.

In Finland, all children under the age of seven have the right to preschool. And in Canada, the province of Quebec offers universal, government-subsidized day care for children ages four and under, at a cost of $7.30 to $20.00 per day. When I spent an evening with two Montreal-based scholars in their twenties, I learned of their city's incredible caregiving prices and understood how this underemployed twosome could already have two children.

Failing the most obvious solution to America's care crisis, and setting aside tax-based measures, perhaps we could make the options that do exist more accessible and easier to find. After all, just finding a day-care spot is hard for most parents. Why not create a national database of vetted day cares, listing pricing and availability? Local versions of such registries exist right now. In the Bay Area, for instance, there is NurtureList, a site that gives parents access, at no charge, to lists of many of the available day cares with spots open in a given area. NurtureList provides descriptions of facilities and fastidious center profiles. In addition to listing day-care openings geographically, the site writes up new schools and specialized programs, like preschools that heavily rely on the great outdoors.

One user of the NurtureList site was Zoe Hanson, who had just moved to San Francisco with her husband and toddler; they moved into an apartment in a "crumbly house" that, at $3,000 a month, was untenably expensive. She didn't have a job and didn't know where to put her child while she looked for one. Her new neighborhood seemed to have "six thousand kids for only three thousand day-care spots," as she put it. Hanson's dilemma was typical of what is actually a national problem. As of 2014, Colorado's licensed day-care spots met the needs of only one-quarter of the state's young children. In Minnesota, the number of in-home child-care providers in three counties declined by more than 17 percent from 2011 to 2016, leading to an extreme shortfall.

It's almost banal to state the terribly obvious: the day-care crisis is ultimately a structural problem. The high price of care is the result of government disinterest. We simply do not have enough affordable child-care facilities.

"I can't even work, as I can't find child care," Hanson told me at the time she started using NurtureList. All she needed was a center nearby that cost $2,000 or less a month. As for so many other parents discussed in this book, her struggle just to find a day-care space for her child was extremely difficult—but resources like the NurtureList database can be hugely helpful. After finding out about her options via this database, Hanson enrolled her daughter in a warm bilingual day care; soon after, she found a job as a studio and marketing coordinator at a design and architecture practice. Of course, like so many middle-class families in this book who are living with the dark shimmer of constantly shifting precariousness, a year or so later Hanson's fortunes had changed yet again for the worse. "I am not working now," Hanson said. She had been working full-time; when she went to part-time hours, her child-care rate fell to $1,200 a month. After giving birth to her second child in June 2017, however, she wasn't working at all: day-care costs were out of reach, no matter how expedient the new-fangled online listing would have made her search for care. "I will not be able to go back to work during the day until my daughter starts kindergarten because I can't afford two in day care. I will be looking into waitressing again in a few weeks when my California parental leave runs out."

One large and crucial solution to limited day-care access would be a national and universal pre-K policy. Locally, this is actually happening, with public pre-Ks implemented in different cities across the country. One of the most effective initiatives now helping hard-pressed parents is New York City's Pre-K for All. It's New York mayor Bill de Blasio's premier achievement, in fact, and its scale is one reason for its success. In 2014, there were only twenty thousand free all-day pre-K seats: two years later, there were seventy thousand. As Dana Goldstein wrote favorably in *The Atlantic*, "At a time of gridlock in Washington, de Blasio has quickly created a new entitlement in the nation's largest city: an extension of the K-12 school system into an additional year of free, academically rigorous public education and childcare. . . .

The program is so popular that suburban legislators have demanded state funding to provide their constituents with the same benefit."

"We know why pre-K is important—the long-term impact on child development, and also the economic impact of paying for preschool on families short-term," one of the architects of the program, Richard Buery, told me. "What happens to a family when they have stable supportive child care? Their stress goes down."

Buery told me that one element of Pre-K for All's success, which other cities and states could replicate, is its "anti–*Field of Dreams* approach," as he put it. What he meant was that the program, instead of taking the attitude of "build it and they will come," very aggressively recruits students. I myself got so many robo-calls and emails asking about my daughter's enrollment that the heavy sales approach began to feel like something out of *Glengarry Glen Ross*. That aggressive outreach was by design, as it turns out. The mayor had staked his career on the care issue, and there was no way he or Buery was going to accept its failure from the children simply not showing up. "We were overcoming language barriers and cultural barriers," Buery said, referring to "the parents who didn't think four-year-olds should go to school."

One lesson of all this was to spend money and personpower up front to recruit the people who will use such programs, with aggressive local outreach that went beyond advertising. New York City avoided letting universal pre-K be like the Earned Income Tax Credit (EITC), which can be so beneficial for low-income families but for years was underused because families simply didn't know about it.

A visit to one New York City public pre-K classroom in my neighborhood was striking. The school was the recipient of special financial assistance due to the high number of children from low-income families. The kids played calmly together under the supervision of a motherly, gray-haired veteran teacher in clogs. After recess was over, they filed into the pre-K classroom and literally learned their ABCs, with surprising vim. I knew that many of the parents of these children lived in housing projects near my home, and I thought of all the mothers whose hard workdays were at the least softened by this new program.

As of this writing, though, only a scattershot of pre-Ks can be found across the country. According to a Hechinger Institute report from 2016, using data from the National Institute for Early Education Research, access to state pre-K varies wildly from state to state: 48.7 percent of four-year-olds in Texas have it, while only around 2 percent of their little counterparts do in Missouri, and roughly 10 percent do in Oregon.

Of course, even in New York City, the pre-K program has no counterpart for the seemingly endless years in a child's life when you are a young parent, especially on snowy days, without a helping hand. The time before age four can indeed seem especially endless when you are living through it. That fact—the total lack of any way to help parents who have babies and toddlers and want to return to work—reveals one of the key failures in how our system neglects families.

AS WE WAIT FOR A MUCH-NEEDED TRANSFORMATION THAT CAN seem most unlikely, the forever work clock keeps ticking by for America's working parents. And that twenty-four-hour punch clock and irregular hours continue to combine with our lack of affordable day care to make many people's lives very difficult. Today, unstable hours are a bigger problem than low pay, advocates told me. However, Joshua Freeman, for one, is optimistic: he believes that the movement for a fair workweek will follow in the successful path of the fight for family leave and a higher minimum wage. The insecurity of work hours is one of those "apple pie" issues: not appearing on the right side of this issue can make corporations look like the clock-loving Grinches they may actually be.

To reset the forever clock, a national movement is under way to create stable schedules via a national pilot program undertaken by the clothing franchise the Gap to phase out on-call scheduling—which, of course, is not exactly the same as creating a consistent schedule for workers. City and state initiatives persuading businesses to shift away from just-in-time scheduling will go a long way toward making American working families more secure. It would help parents who currently can't be home from work in time to help their kids with their homework, and sometimes not even in time to put them to bed. And it will be a crucial development when companies and organizations institute fair hours, not just for McDonald's workers but also for adjunct professors.

"It was all about nine-to-five day care ten years ago," said Deloris Hogan, director of Dee's Tots. "But now that the stores are open till twelve at night or even twenty-four hours a day, we are needed. We have to do this."

Soon somebody, if not Deloris Hogan, will open a business to accommodate the 3:30 A.M. drop-offs. Take Kaden, one of the children at Dee's Tots. Kaden's mother is a single mom who works at Costco until 10:30 P.M. Two-year-old Kaden was so well behaved that he washed his hands at the day care's play-kitchen sink before cooking his *play* food. Many days, Kaden's aunt picked him up at another day-care center and dropped him at Dee's. Even tiny Kaden was living on the forever clock.

One reason for the forever clock is the rise of single-parent households: in 2013, 28 percent of children were living with a single parent, and 77 percent of those single parents were mothers. Even mothers who have the support of family and friends may strain those relationships to the breaking point. When single parents are just scraping by, scholars have found, they rely greatly on their closest relatives for support—far more so than their wealthier counterparts. And while such reliance can breed intimacy, being more reliant on your friends and family (by 30 percent, according to one scholar I spoke to about single parents) can also backfire. A single mother who relies too much on her neighbors or her sister can't offer them much in return and thus often strains those relationships.

The state paid the Hogans up to $250 per child per week for families who qualified for assistance. A few families could afford to pay that rate in full themselves, but according to Patrick Hogan, most paid only $1 to $5 a day, with the state making up the rest. It is not easy to work with the social welfare system, which is a site of the culture wars between the privileged and the lacking and thus under social pressure and endangered.

On one of my visits, giant puffy pencils and teddy-bear balloons were hanging from the front fence and kids were jumping on a bouncy house, giving the place a touch of the carnival. At twilight, two children helped Deloris water her cucumber and melon plants while keeping their eyes out for plundering raccoons. Dee's Tots did not focus on education. The Hogans worked a bit with preschoolers on the alphabet and numbers, but the children's hours there were typically consumed with playing with dolls and their dresses, throwing and kicking plastic balls, engaging in sing-alongs, or dancing to R&B hits or "Cha Cha Slide."

These twenty-four-hour kids seemed all right, for now, at

least, and unusually self-sufficient. One night, Diana, Marisol's eight-year-old daughter, showed me how to perfectly squeeze toothpaste out of the tube, like in a commercial. Her four-year-old sister, Ivette, brushed her own teeth without being asked. I saw little of the sorrow or loneliness one might expect to see in these kids, though I did catch ripples of uncertainty and need. A few of the children did seem "parentified," in the useful but awkward psychology parlance—they seemed wise beyond their years, having been forced to grow up and parent themselves earlier than we might wish.

Later, the Hogans played the film *Coraline,* a dark fantasy about a child whose real mother, in a parallel world, is substituted by a fake mom, complete with buttons for eyes. One of the girls, two years older than my daughter, turned to me and said, "Why do children think parents are going to save them?" She paused. "That's stupid."

4

OUTCLASSED

LIFE AT THE BOTTOM OF THE TOP

On the face of it, Shaun Tanner wasn't someone who needed saving.

By most measures, Tanner, at thirty-eight, made a fine living, in the very low six figures. He had a $3,000 monthly mortgage, which was less than half of his monthly income. In 2016, he worked three jobs to pay for his health insurance and his family's expenses: as a meteorologist for Weather Underground, a weather site; as an independent contractor; and as an instructor at San Jose State University. In other words, he was not as under siege financially or as pressed for time as the Hogans, who ran the twenty-four-hour day-care center, or the frantic parents of the children who attended it.

Nevertheless, Tanner felt tense. Living in the shadow of the tech industry had been bad for some of his friends too: most of them didn't have "ridiculous" tech jobs, but the cost of living created alienating comparisons.

Tanner worked hard without reaching the absurd level of pay of his tech neighbors. His jobs entailed a one-hour commute, each way, between his home in San Jose, California, and Alameda. He exerted himself in order to pay for after-school programs and day care for his three kids, the oldest of whom was ten.

"People see I work in the tech industry and they think I should be a multimillionaire," Tanner explained. He was not. "It's getting worse, by the way, around San Francisco. Should I pack it in? What am I supposed to do? Mr. and Ms. Jones of the Bay Area ask: 'Am I doing as well as my neighbors?'"

I thought I should take a look at people like Tanner, those who were distinctly upper-middle-class but who were still running in place in order to stay that way. I did so because I knew they too felt an emotional and to a smaller extent a financial sting as a side effect of the extremes of income inequality in this country. "Week to week," Tanner said of his life, he "still feels a crunch" on Fridays. "There's nothing left here."

The complaints of the relatively privileged can raise hackles. Their problems can seem bespoke and psychological problems akin to memory exercises, specialized summer camps, or homemade jam. The stresses felt by quasi-privileged people contending with income disparity may *seem* less real than those suffered by the many. But they are real in their own way, and they also reveal much about the concentration of extreme wealth in some American cities, and the gaps between the one percent and everyone else.

Upper-middle-class American life today stands on frail foundations. Psychological and social science research demonstrates that living amid the wealthy even when you are reasonably well salaried yourself is damaging to your mental health. In 2010, a study by researchers at the University of Warwick and Cardiff University found that money improves happiness only if it also improves people's social rank.

In other words, being highly paid isn't enough: people want to see progress in their lives, to feel as if they are moving up, and to be able to exhibit that ascension to people in their community and to themselves. Glenn Firebaugh and Matthew Schroeder have also written on this in a study entitled "Does Your Neighbor's Income Affect Your Happiness?" "With respect to income and happiness," they write, "what matters most is how much income a person has relative to his or her income comparison group." Personal contentment is relative: it depends on how you see yourself compared to your "genuine" peers.

Like Tanner, Amy, who had worked in human resources for over eighteen years for different tech companies, had contended with the assumption that anyone working in and around Silicon Valley "should be raking in six figures." But it wasn't the case in her line of work. Amy and her husband had a combined annual income of $150,000, which sounded like a lot, she acknowledged, adding that "anywhere else in the country that would be a ton of money." But they spent over half of their monthly income on their mortgage; child care ate up another 30 percent. Amy also spent a large amount of time commuting because she couldn't afford to live closer to her job. Every day, "I literally drove my kids out at 6:00 A.M., to get them to day care by 7:00, get into the office at 8:30, pick them up at 5:00 or 6:00, then we'd commute back, then bath, and food, and bed. I'd spend fifteen minutes a day relaxing with them." As she put it: "I didn't make enough to get the cleaning lady or the chef like my colleagues did—my colleagues didn't have time to clean for themselves or feed themselves, but they did have money."

Scholars now research the need to understand the local perception of wealth, poverty, and status and how it affects people like Amy. For instance, danah boyd (she prefers to style her name in lowercase letters), founder of the think tank Data & Society, argues that social status may be local. As boyd points out: "People's understanding of prosperity is shaped by what they see around them."

While Americans overall may live better than medieval aristocrats could even dream of, that means nothing when oligarchs live next door, flaunting their luxurious homes. If you are upper-middle-class according to national averages yet dwell in the hyper-wealthy areas of this country (all the richer thanks to inequality), you may not feel your privilege. In my interviews, I heard how isolated these parents felt. I couldn't help but suppose that their sense of anomie was at least partly derived from intraparental competition—a solitude only heightened by dependence on social media, with its Facebook-boasting surfaces, as a conduit to the lives of others.

I recognized this class-based feeling of parental alienation in my own life. Living in wealthy cities, one can feel like one is being dominated by the rich. When I write the dastardly word "domination" here, I do not use the word in either the sultry or the not-safe-for-work sense. It is also what we feel around those with financial capital when we have less of the same. It characterized a feeling I had of being corralled into certain corners of one's home city, owing to limited income and holdings. It was also the suspicion that I was paying "rent" every time I went out for coffee or a walk, that there was no longer any public space to sit in, that a high charge was always associated with simply "hanging around": once again I'd have to buy that unwanted second overpriced spice tea, or hand over another chunk of change for an hour at the indoor play space for my daughter.

These were the small degradations of my middle-class daily existence in an overpriced metropolis. Individually, these pinpricks were bearable, but together they combined so that the city seemed to whisper to my kin and myself: *This is not the place for your kind anymore.* (My husband and I had scouted out other, cheaper cities, like Austin, Texas, and Portland, Oregon, in the hope that we could easily start anew in one of these places, collecting vintage clothing covered in peacock feathers and eking out enough of a living to just barely afford nitro iced coffee. But we eventually chickened out and stayed where we were—the fiendishly overpriced place that quite simply was the one with jobs in journalism.) At the same time, I started to see stores in my neighborhood close almost as soon as they opened because small shop owners could not pay their rent. They had been priced out of their own shops, and the blocks near my home immediately resembled a small ghost town within the now-imperial American city where I lived. Indeed, as for Tanner in San Francisco, my city and my spaces and the schools I attended as a child had been invaded by the ultra-rich, and that was one reason why rents were so high. If the merely middle-class reject the idea that the upper-middle-class also suffer, they deny solidarity with another group that *is* in fact being harmed by inequality, no matter how gently.

Again, how your income stacks up against the income of others in your surrounding community is more crucial than your pay itself to your satisfaction—both mental and physical, according to another 2014 study. Michael Daly, an associate professor in behavioral science at the University of Stirling, and his colleagues write that their findings "demonstrate that social position rather than material conditions may explain the impact of money on human health." In the study, the rank of a person's income and wealth, rather than the absolute of these things, was a better predictor of overall health outcome, including obesity and chronic illness.

Meanwhile, having a high standing had actual health benefits. To put it in social class terms, it was better to be at the top of the bottom economically, or the top of the middle, than at the bottom of the top. Being placed at the bottom of the top, it seemed, had a corrosive effect on physical health.

On the face of it, a good number of people ranked at the bottom of the top are doing fine economically. But they can still feel besieged. Even some of those who are holding on professionally and seem safe financially are in emotional trouble.

I must reiterate here that of course the psychological burden on the upper-middle class isn't anywhere near as dangerous or even as anxiety-producing as it is for the less solvent families you have met. But there is a specific pain felt by people like Tanner, and human resources workers like Amy, who live or lived in American cities now dominated by the 1 percent. I feel it too. I recall New York City before it was gilded, when it was still in its 1970s and '80s era of detritus, the setting for the popular pose of anti-ambition. Thinking of this time before the real estate boom in my city fills me with what I think of as dirty nostalgia. Back then, the adults I knew and loved lived in slovenly rooms that were strung together like beaded necklaces encrusted in grime. Now inhabitants of my neighborhood are the truly rich: for many millions of dollars, they purchase brownstones that the novelist Edith Wharton once described disdainfully as homes coated in "a cold chocolate sauce." I watch as these homeowners—or more usually their nannies—arrive at day care and school to pick up their children wearing white jeans and spotless sneakers, which they often also wear (incongruously) to the playground, usually carrying giant, four-figure designer handbags with clanking ornamental clasps. The cleanliness and the interest in disposable yet costly objects consistently made me nervous, clothes as literal white flags signaling a seemingly limitless earning capacity (mostly acquired through financial activities that almost certainly heightened inequality). The wildly expensive children's birthday parties they throw in their new marmoreal apartments—with sometimes a few party planners on hand and multiple day-care providers—seem like surrealist exercises.

In the past, those in the upper-middle class weren't all trying to keep up with the 1 percent or with celebrities. They also were likelier to have a professional certainty and stability that made up a little for their lower rank. No longer is that true.

Both readers and trolls commenting on media sites reliably calling out upper-middle-class people who complain about this state of affairs, heckling them for their relative comfort and privilege. I am looking at you, Neal Gabler, who in 2016 was hated on for his *Atlantic* essay about his own supposed financial distress, which included the sad tale of living in the wealthy New York City vacation spot the Hamptons and cashing in his retirement account to pay for his daughter's wedding. In response, the writer Helaine Olen described the genre of "sad, broke, literary men": "a successful, well-respected, white male author steps forward to air his money woes, a tale of kept-up appearances and secretly abysmal finances. He is, he says implicitly or outright, speaking for all of us. . . . But all too often, it is simply a disguised narrative of privilege."

And sure, these commenters (and I) might be dismissed as petit bourgeois kvetches. Going in and out of your city's proverbial "poor door"—the separate entrance for income-restricted residents of mixed-income housing—every day has its psychic costs. And again, the toll remains hidden because people who are fiscally outranked are all too aware of not only others' but also their own in expensive cities like San Francisco, where the median rent for an average two-bedroom apartment runs tenants $4,430 a month.

The stories I heard from upper-middle-class people who live next to the warped and wealthy included this one: In a Maryland suburb, a middle-class high school varsity baseball team was playing against a team from a wealthier suburb nearby. A mother of a player from the not-so-wealthy team was happily cheering on her son and his teammates when she heard the "cheer" of the other team, issued by both adults and kids: "Lower-average income! Lower-average income!" And then, "Can't your parents afford to feed you? Can we call child protective services?" These were 1 percent jokes, coming out of the mouths of the very young.

Research conducted, beginning in 1967, on British male civil servants by Sir Michael Marmot for a famous document known as the Whitehall Studies further demonstrated that income inequality has mental and physical health effects, even among executives who are one rank below top-level administrators. In the study, the very lower-upper-class cohort had higher mortality rates than their bosses: of course, that mortality rate only got worse going down the scale to the most stressed employees at the bottom. But the gap between people at the very top and at the bottom of the top was striking. Such gaps are the result of overall inequality: even those in the top tenth percentile of salary in the United States now see themselves as excluded from American power and wealth, as studies have shown. One such study, a 2014 Brookings report, found that two-thirds of the one-third of Americans who live paycheck to paycheck might be considered well off. The authors of this study, entitled "The Wealthy Hand-to-Mouth," describe a group with an enviable array of illiquid assets like houses and retirement accounts who nonetheless see themselves as financially struggling.

Upper-middle-class families also are daily engaging in what economists at the University of California, Berkeley, and the University of Chicago have called "trickle-down consumption": spending money to hire tutors to help their children compete against the progeny of the very rich, who are getting even more extra help for their kids to be academically and athletically competitive—spending, for example, from $65 to $120 per hour on math instructors all the way up to $500 an hour.

These are the emotional results of income inequality. But stress and invidious comparisons are not just in these people's privileged, neurotic heads. Why? A number of foundationally middle-class professions have been quietly going to seed. Like law.

FOR YEARS I HAVE BEEN MEETING NOT ONLY DISCONCERTED AND underemployed lawyers but also coders, technology librarians, and human resource workers. I ask them why they are financially unstable, and they answer eagerly, as if they have long waited for someone to listen to them. Yet paradoxically, they tend to want to stay anonymous. I suppose that talking about what went wrong when they took the straight-and-narrow path would be the professional equivalent of a walk of shame. And what if their practices took off and anyone could read their admissions of prior failure in a book?

I talked, for instance, with a Southern California lawyer in a private practice who worked on personal injury cases and civil litigation. Despite working long hours, this thirtysomething lawyer and his wife had to live with her parents to save on rent. His immigrant parents were eager for grandchildren, but he couldn't afford to have children, he said. He and his wife simply didn't make enough, and his debt from law school still lingered.

Stories like this are common. Ainsley Stapleton, a forty-year-old accountant based in Arlington, Virginia, who described herself as middle-class, has three children. When I first met her, they were all in either preschool or day care, and she calculated that she spent 87.6 percent of her take-home pay on their care.

"It makes me want to cry a little," Stapleton said by phone from her office. In the past, she said, she and her husband, a government employee, had discussed whether "he should quit or I should, but both of us enjoy working."

When I spoke to Stapleton again a year later, things had improved. Yet some unexpected—but nationally typical—pressures remained. With only one child in day care and two children in a public elementary school, the family's overall expenses had gone down. (Her third child entered kindergarten in September 2017.) But now there was the additional cost of summer camp and after-school care for her two older kids. She would have to work past three in the afternoon most days and for much of the summer. In other words, they had to enroll their kids in the camps and sign them up for more day-care hours if she was to do her job properly. Northern Virginia summer camps weren't cheap, she complained. There were cheaper camps, but the average cost was still probably around $350, she estimated; the camps her children went to cost $425 to $450 a week per child. For Stapleton, these unexpected new costs had to be added to the $21,000 the family already spent each year on day care and infant care.

Costs such as these, which can take a big chunk out of a family's take-home pay, are highly variable depending on geography: according to the basic budget threshold set by the Economic Policy Institute (EPI) for a two-parent, two-child family, in 2015 this unit needed to bring in $106,493 a year in Washington, D.C., while the same family would have needed only $49,114 in Morristown, Tennessee. (The EPI also found that child-care costs for a family with a four-year-old and an eight-year-old exceeded the average rent in 500 of the 618 U.S. communities it surveyed.)

On the surface, Stapleton's family is not unhappy, she said. They have a house, with a mortgage, and in nice weather they spend hours at the park on weekends. But her children's childhoods are different from her own outside of Philadelphia. Back then, her mom stayed home with her and she even went to private school; now, Stapleton knows, extras like these are outside of her family's reach. So is going on vacation. Stapleton is a CPA who "calculates numbers all day for other people," as she put it, so it seemed strange to me that she didn't know the details of her family's expenses offhand. She admitted that she didn't want to do her own numbers because she just couldn't bear to look at the results.

As I spoke to many other middle-class people who admitted to being even more economically wobbly, I wondered the degree to which they were authors of their own disappointment. Some were melancholy. Sometimes it seemed both personal and due to larger external factors, like ageism. Take Anne, a lawyer in Mississippi, who described her life bluntly as "not a happy one." At fifty-nine, she had worked for over twenty years at a firm specializing in tax law. When she started having migraines in 2010, she went on three months of medical leave, but then was eventually "pushed out" by her employer, she said. She depleted all of her savings after she became bedridden with her chronic headaches. At the same time she took time off for health reasons, she was paying for her son's college tuition. The result was grim. "This is a vastly different life from what I expected to be having at this age," she told me. "The six-figure salaries and benefits are long gone." Her by-now-familiar (at least to me) and biting words about law schools and bedeviled firms followed. There was also the woman who was working to keep up appearances in Manhattan, trying to seem appropriately positioned and similar to her peers by, for example, spending on little luxuries—"ordering a glass of wine rather than just drinking water and using a credit card more often to pay for, say, hair color or highlights." She added: "I think of myself as *poor bourgeois*."

Michelle Belmont, whom we met in the first chapter, the library technologist, stayed in school studying for a master's unrelated to her profession just so she could defer paying back her school loans. At the time, in 2014, she was paying off her Visa card with her Amex card with her MasterCard. "I feel like we are poor but we are supposed to be middle-class," Belmont said. She was trying to figure out what that meant and why she couldn't survive in today's marketplace. She could barely pay for her son's day care, which, at $245 a week, was the cheapest she could find. Again, like other squeezed parents, she blamed herself, not the broader social system. "Maybe I could be upper-middle-class without the debt?" Belmont wondered. She'd had a credit card since she was eighteen and had racked up years of what she considered unnecessary debt. "That was stupid," she said.

I also had a conversation with yet another mother who ran a small business and was nearly broke. Her child was on a nearly full scholarship at a private school. (She requested anonymity, ostensibly to protect her child's identity.) She couldn't afford to buy her daughter the basic things other kids had—such as the expensive sweaters that were part of her daughter's school uniform. And she sometimes missed school and parent meetings, which were consistently held during the workday, and even some of the evening meetings when she couldn't afford the extra babysitting cost.

"I tell my daughter that some people have a lot of money and have no love, and some people have no money and a lot of love," she said. "It has forced a conversation before it would have happened otherwise, but in wealthy and unequal areas it has to happen early. You have to keep the kid strong and confident in themselves through this."

This mother's words may sound like hard lessons but it is, I have found, a nearly impossible conversation to avoid having with a child growing up in a city like New York, as the hysterically affluent were constantly in these children's faces. These extremes were numerically true as well. As Emmanuel Saez, an economist at the University of California, Berkeley, has famously written, between 2009 and 2015, 52 percent of real income growth flowed to the 1 percent. (The Equality of Opportunity Project has found that correcting even slightly our country's unequal distribution of wealth would help many more children replicate their parents' success.)

I also spoke with a family of middle-aged professionals in Florida beset by student debt. Robert Mara entered law school believing that he was training for a profession that would guarantee him a spot among the securely upper-middle class. Ten years after attending the for-profit Florida Coastal School of Law, he and his wife, Wendy, whom he had met at Florida Coastal, had two kids, a house with a mortgage, and five degrees between them, three of them advanced. Robert had served in the military and managed a supermarket before he started law school. Now in her fifties, his wife ran her own legal office near Daytona Beach. Robert shared it with her for a while, but then got fed up with having a legal practice. He disliked the hustling, and the practice was not as viable financially as he had hoped. So he embarked on a third field, finance, and left the struggle to maintain the practice to Wendy. Then in his early forties, Robert was receiving an entry-level paycheck of $36,000 per year. The Maras were also $400,000 in debt for the cost of their educations. That they were unlikely to ever be able to pay it all off, Wendy said, made her want to cry.

These lawyers' complaints about economic stress should be one-offs, the exceptions that prove the rule, but they aren't. This is due to the downward direction their previously secure profession has taken. After the 2008 recession, law firms and corporations retained fewer lawyers, and there was also the problem of student debt. It had risen from $95,000 at the average law school in 2010 to roughly $112,000 in 2014. According to Matt Leichter, an attorney and writer in Minneapolis, legal unemployment is also worse in certain states: in Alaska, 56.7 percent of those with a law degree are not working as lawyers. In contradiction to the myth of the hardworking southern lawyer propagated by the legal novelist John Grisham, in Tennessee only 53.6 percent of those with a law degree work as lawyers; in Missouri it's 50.8 percent, and in Maryland it's 50.3 percent. Meanwhile, Leichter has calculated, there are "excess" attorneys (per ten thousand residents) in all but three states. Leichter runs a site promisingly titled The Last Gen X American, which, he notes, was a "better use of my time than playing *Tetris* while listening to garage rock all day."

This was not what I and so many others were raised to expect of the legal profession. "Lawyer" has for more than a century been a code word not only for "the Establishment" but for security. In the 1970s bestseller *The Stepford Wives*, for instance, the housewife Joanna is married to Walter, the ultimate "boring lawyer," and lives in the suburbs. That was the idea of the lawyer back then—someone traveling the dull but safe, predictable, and reliably remunerative path, one that literary types like *Stepford*'s scribe Ira Levin or, say, Philip Roth were most likely trying to avoid. Now "lawyer" as monotonous and reliable earner seems a dated curio.

Now, as the *New York Times* reports, ten months after graduation only 60 percent of the law school class of 2014 had found full-time jobs with longtime prospects. Data have shown a decline in hiring by law firms as well as the government. Thousands upon thousands of recent law school graduates have been caught in this downturn, which resembles what has happened with some of the other middle-class professions we have read about here. And just getting a legal education winds up burdening lawyers with debt: squeezed America is often indebted America. In the last decade, tuition debt in general has quadrupled. College costs have risen more than 1,000 percent since 1978, and what American students and graduates owe overall has surpassed $1.3 trillion. Moreover, it is not held just by college students—their parents have taken on debt as well.

It might seem coincidental that these rising levels of debt and fading opportunities are happening at the same time that college and, to a lesser extent, graduate school access has been democratized. But this paradox seems like no accident to me.

In fact, the "diploma mill" for-profit schools and colleges remind me of the insights in Didier Eribon's 2009 memoir, *Returning to Reims,* about growing up a poor French son of the provinces. Eribon was not supposed to go to college, let alone become a famed scholar. As an ambitious and bright teen, he, like other young people "from less advantaged classes," as he put it, didn't understand the "hierarchical nature" of educational institutions. Such young aspirants, writes Eribon, go to colleges or graduate schools that may be almost useless to them practically. They enroll in programs in the belief "that they are gaining access to what has previously been denied to them, whereas in reality, once they have that access, it turns out to mean very little, because the system has evolved and the important and valuable place to be has now shifted somewhere else." In other words, it's not an accident that these working- and lower-middle-class students wind up in the French equivalent of the less-than-effective graduate schools or law schools in this country, while their wealthy counterparts enroll in schools that may actually help them. The born-wealthy tend to understand that education is a strategy. They would know not to enroll in a school with a spoiled reputation, and they would never join a specialty of a trade that has run its course. Here Eribon puts his finger on what seems so true and horrible to me—the formerly male-dominated professions that have become predominantly female are suddenly less desirable: once feminized, a profession's pay stagnates.

The system is rigged against people who lack exclusive social knowledge in yet another way: the value of the best and most desirable institutions or jobs is mutable, in the sense that graduating students need to somehow know what the next best move to make is, drawing on their social capital. Those who have no such knowledge will lose out.

Robert Mara saw this lack of knowledge about various degrees and their professional possibilities at his for-profit law school, Florida Coastal School of Law. "There's not enough work in law, so it is very cutthroat," he told me. "If you weren't top of your class, or didn't go to the right law school, there were just simply too many attorneys out there. My law school was a degree factory." Mara described jammed classes and students who entered school with very low LSAT scores. These were some of the future "excess attorneys" whose reach exceeded their grasp.

The experience of fifty-year-old Kiki Grossman, who did administrative work in an accounting department before she attended Florida Coastal School of Law in her late thirties, bears this out. A shortage of legal jobs has left Grossman without a car or a home, and she and her husband live with her mother. Unemployed, she filed for Chapter 13 in 2014, when she was also still deeply in school debt. (As of 2017, she had a new job, at Valencia College Peace and Justice Institute, where she served as the coordinator for the Legal Education Action Project. She was still plagued by student debt.) Grossman didn't blame for-profit law schools, however, for her misfortune. The "paradigm" for legal work had shifted, she wrote me in an email. "The school can't be blamed for bad timing."

"Where I think the school screwed up was in its financial counseling," said Grossman. She hadn't realized that she could take out federal loans, including cost-of-living loans. "The private loans were presented to me as my only option."

Among all law schools, Grossman's law school had one of the lowest median LSAT scores, at 144. Tuition there was expensive, at $46,068, and as of this writing a very low percentage of graduates—61.9 percent—are employed as lawyers. Many of these students would be unlikely to pass the bar, she surmised—it was, as the legal higher education critic Paul Campos wrote of Florida Coastal in *The Atlantic*, "easy to make the case that these students wind up in far worse shape than defaulting homeowners do."

Each one of these "legal cases," so to speak, of thwarted lawyers might seem anecdotal or, even worse, created by a society of victimhood in which personal malingering is projected back onto society as a wrong. *You fail and America is to blame!* But when I checked into it, I discovered that, according to Joe Patrice, an editor of the legal journalism site Above the Law, lawyers may indeed be making one-quarter of what they were making before 2008. "Even that would be a decent salary if they weren't in debt from law school in the first place," Patrice mused. "Their pay is also stagnant."

One of the problems, of course, is that law is not unionized. It's a highly individualistic profession, Patrice noted. As in so many other middle-class professions and working-class jobs, the lack of solidarity takes a toll. In addition, in law there is the looming threat of automation: paralegals and legal assistants face a 94 percent probability of their jobs being computerized in the future.

As with the academics you read of earlier who are barely surviving, the plight of lawyers highlights that pursuing even some of the most "respectable" professions can lead to career stagnation and to being part of the uncomfortable, unstable middle class.

We could blame the underemployed lawyers themselves, but we could also see them as treading the latest dividing line between the dominant and the dominated in America. The relation between the privileged and the less privileged "reproduces itself by changing location," writes the sociologist Eribon. That's how the so-called democratization of certain professions or schools, such as law, "is really a displacement" of the previous ways in which some people were powerful and others were less so. Inequality then expresses itself more finely. It is this subtle process of displacement, Eribon suggests, that keeps the structure of inequality intact.

It is also the kind of thing that has driven law school "reform and transparency" activists like Kyle McEntee so mad. McEntee, a law school graduate now in his early thirties, cofounded a nonprofit called Law School Transparency.

He believed that he had a fresh riposte to the lawyers' conundrum. "Law school is a very, very unfortunate decision," he began in our first conversation. Law schools don't always offer full information on job or salary outcomes. At least some employment statistics are required by the American Bar Association, but while some schools post more than the minimum—voluntarily disclosing employment and salary information—many do not. On top of all that, though the very cheapest law schools cost a little over $10,000 a year, the vast majority charge tuition in the range of $40,000 and above. Law School Transparency's official mission is to provide information so that people can make informed decisions, a kind of consumer protection. McEntee said that he sought to empower those going to law school financially so they would understand what lay ahead. Prospective students should at the very least negotiate down the amount of their tuition; McEntee asserted that he had done so himself and that it is very possible to obtain what he called "tuition discounts." This is all part of Law School Transparency's project of putting law schools on notice about recruiting and admitting students who are likely to be burdened by educational debt for the rest of their lives.

To advance this project, McEntee helped conduct a study in 2016 with Deborah Merritt, a law professor at Ohio State University, which found that while women are nearly half of the student population at American law schools—preparing for what was previously a male-dominated profession—women are nonetheless more likely to attend lower-ranked schools than men and thus are less likely to get sustainable, well-paying legal jobs. The study was called, pungently, "The Leaky Pipeline for Women Entering the Legal Profession."

While McEntee may be something of an outlier, his movement is doing something to pierce false consciousness:

he is trying to reframe law school and even the icon of "the lawyer." This shift in consciousness could be broader still, raising new questions about the merits of getting a college education at all as a growing number of people question the benefits versus the cost. A related solution is to shorten the length of a law school education to two years from the standard three years (President Obama suggested this change too) and also to offer practical concentrations, such as in contracts.

McEntee was personally angry about his experience with law school. Post–law school, he was unable to buy a house, and he didn't think he would be able to afford to get married or have children. In fact, his student debt had led him to delay all sorts of decisions. "Law schools chew us up and spit us out," he said when I first contacted him a few years ago. "We are supposed to be the leaders going forward in the country, and so-called successful people, but instead I am eating ramen." (In 2017, he said that he still thought he would have difficulty buying a house, but then didn't want to comment any further on his individual situation.) Law school could be said to have once been a pathway to power. It still is for some, but now, if you go to the "wrong" law school or live and work in the "wrong" state, the equation doesn't quite work. The white-collar witches' brew of a degree plus years of work in the field no longer boils and bubbles into status and traditional success. The people who believed that it would may now be smarting from a broken narrative.

Another fix for disgruntled and sometimes disgusted lawyers is—no surprise—self-help and specialized counseling. Lawyers in San Francisco can obtain a mental coach via an organization with a caustically hopeful name, Leave Law Behind. "There is an easier, less painful, less stressful, and lucrative way to make money," its website declares. Leave

Law Behind's Casey Berman, a fortysomething former lawyer, told me that he saw his mission as "motivating" former lawyers who are either broke or deeply frustrated, or both, and getting them to do something else with their lives.

The basic truth is that like other squeezed families in this book—and like all social groups in general—the upper-middle-class people you have just read about are trying to reproduce their own status. One theory of social class from Marx onwards is this: in their lives, people are simply trying to replicate their own status position for their children and to cement their own class legacy. A social class, however, can be all too easy to drop out of, especially today.

As the journalist Tim Noah puts it, the top 10 percent of Americans are "sort of rich," the top 1 percent is "rich," and the 0.1 percent is "stinking rich." The spectacular success of the 0.1 percent, a tiny portion of society, shows just how stranded, stagnant, and impotent the current social system has made the middle class—even the 10 percent who are upper-middle class. Those "excess attorneys" I spoke with will not be passing on their social class standing to their children. And those in the precarious-feeling upper-middle class in the wealthiest cities may be thwarted in their attempts to, for example, own property, all the while living amid astounding levels of affluence.

Work and class identity may be games, as some have argued. But they are games that we can't win or reliably win, at least not anymore. Shaun Tanner, the tech meteorologist, wanted to quit playing this game. He longed to escape from his dominated and stressed upper-middle-class position and neighborhood. Maybe go minimalist? Move to the forest? "Live off the land," as he put it? I have that wish too. I also recall New York City and San Francisco and L.A. when they were ruddier, more bohemian, and less 1 percent, when they were a palladium of the thoughtful. But how can I leave my opulent city? After all, cities like New York are now the places that, paradoxically, provide workers with jobs in tech and media and law. Opportunities to practice these classic professions have waned in the noncoastal geographies.

Thus, a sliver of America remains stuck in the middle, keeping up appearances for others and for ourselves. At least, for as long as we can.

5

THE NANNY'S STRUGGLE

Blanca waited for Guido in the metallic airport lounge. When would he come? An hour passed. Then another. Blanca made three different trips to the arrival board, ticking off the flights. She checked her texts again and again. Was her son here? She tapped a note in Spanish to her best friend Gloria, who had accompanied Guido on the fourteen-hour flight from Paraguay. In response, silence.

Blanca would have to wait. But she was used to this. It had been ten years since she lived in Paraguay with Guido, who was eleven years old. She had left him when she was in her early thirties and ever since then had supported him and her aged mother with her wages sent back from working in America as a nanny. When she left, Guido had resembled the chubby toddler with thick little legs whom she nannied in Manhattan.

"*Estoy nerviosa,*" said Blanca. *I'm nervous.* Finally, the tension broke as she got a text. They had landed. "That means they are in Immigration. So close, so very close."

Blanca was hardly the only working mother living across the globe from her child. Academics describe such arrangements as links in the "global care chain." At one end of these links is a woman in a developed country. She gets a job and is unable to take care of her own children full-time. She hires a low-wage worker from overseas. These immigrant nannies, in turn, hire caretakers back home who earn even lower wages. The monetary value of women's labor declines as one follows the chain from the Global North to the Global South. The chain works by separating wage-earners from their dependents. University researchers studying Latina immigrants in Los Angeles estimated that 24 percent of housekeepers and 82 percent of live-in nannies had left kids behind. Blanca's was a more hidden variation of the squeezed parent stories you have read about so far. Unlike those who must put their children in extreme day care and the day-care workers who overwork to serve them, Blanca was both the most squeezed of parents *and* day-care providers—so pressed that she had to leave her son for a decade back home, where he grew up without her.

Now Blanca had to make another hard choice: whether to pay the price of getting her son back by replacing his middle-class life in Paraguay with becoming part of the working poor in America.

This hasn't always been the only choice: as the Columbia University historian Alice Kessler-Harris, author of *Women Have Always Worked,* tells it, at the turn of the last century and at different points in the twentieth century, America was indeed a land of opportunity. Immigration was always difficult, but it could be a pathway to success. Now, says Kessler-Harris, there is less social mobility in the United States than in most industrialized countries. According to studies, upward social mobility may be as difficult in the United States as it is in Britain, a country famed for its ironclad social class structure. This is counterintuitive, of course. We've always thought of America as a socially flexible place, especially since the 1950s, when wages were generally increasing. But now, with paychecks barely budging, workers like Blanca who are trying to support a kid on one income may get stuck, says Kessler-Harris, as two incomes are now a necessary ingredient in the recipe for middle-class success.

The lack of mobility is best seen in American places like the Mississippi Delta, where there is less social movement statistically than in the rest of the developed world as a whole. (Of course, as Mississippi has one of the largest percentages of black citizens of any state, one can easily say that lack of mobility is not just about economics but where it meets up with racial discrimination.) Among children who might be categorized as working-class or lower-middle-class, the likelihood that they will move up to the top quintile has fallen significantly. The poor tend to remain poor, having simply made the mistake of being born to the wrong parents, as the political theorist Michael Harrington would have said, or born to the "wrong" race, as upward mobility is more expected, and far more likely, for whites. But for the middle class, the largest group of working people in the United States, stagnation is experienced as a great loss, as the end of the mobility and flexibility we saw in our parents' lives. That mobility is so much a part of the American promise that losing it seems like a deep betrayal.

In her native Paraguay, Blanca worked as a nurse in Asunción, but here that class perch had eluded her. And if it was this hard for a legal citizen like Blanca, imagine the lot of those who are undocumented.

Blanca was a woman laboring hard and in good faith to better her chances and her son's. An aspirant to the middle class, she was having a hard time entering it. Blanca's story was about access. Among many other things, being middle-class is a matter of having access to certain goods and services. It's not just the house or the car you can buy. This status is also more granular, reflecting refined varieties of knowledge and information: the middle class knows where to shop for food or find a home, where to send their children to school, where to get medical treatment, child care, career advice or training, or other kinds of help. Perhaps most importantly, class status is about how you even find out about these things to begin with, which again brings us to "cultural capital."

When I recall "cultural capital," I think of my favorite theorist from when I was a graduate student, Pierre Bourdieu. Bourdieu theorized that capital extends beyond economics, encompassing credentials, skills, and tastes. Financial capital is convertible—if you have the latter, you can gain cultural capital through education. Then, if you have the former, you can convert that back into even more economic capital through the right social networks. While some cultural capital is evident through, say, your vintage LP collection, the ten Louis Vuitton bags in your closet, or your Prius, cultural capital is also reflected in an advanced degree or, in Blanca's case, an ability to choose and obtain access to a "good" school for your children. Cultural capital is less available to people in positions like Blanca's than it was in the 1910s or the 1950s, said Kessler-Harris. "If Blanca had migrated in the 1950s, her son would have been likelier to do better," she noted. "There was a greater faith in public schools, greater mobility for a kid than for their immigrant parent: he could have helped to support her by working his way through college."

In addition, now more than in the 1950s and '60s, San

Francisco and New York City and their suburbs are predominantly class segregated.

Blanca's story is about barriers to entry into the middle class and the gaps in the systems that serve and support middle-class aspirants. Her story, inevitably, is about the pervasiveness of class privilege.

We have met parents who were squeezed by pregnancy or early maternity. We have seen how parents' work hours have squeezed the time they can spend with their children, and how the cost of day care has forced these parents to work even longer and harder to afford it. With Blanca, one can see how both culture and economics are failing to replicate the historical middle class—the middle class of immigrants who worked hard and moved up in the world. These parents are now failing in their efforts to replicate the American Dream of being part of the middle class.

Blanca lived in the underlit corners of gentrified and opulent New York City, with its glass-encased luxury towers and iridescent shops. Her very modest take-home pay at the time was about $30,000 a year, out of which she paid for her own health insurance. She also sent much of what she earned as a nanny, along with the approximately $100 a day she earned cleaning houses on weekends, via Western Union to her family back home.

"Supporting my family makes me happy," Blanca told me on a wintry night after work. "But it's always about my son, my father, my mother. Nothing is about me." While Blanca slept in a cold apartment with only two small windows and lived only one lost job away from poverty, she had been able to send her son to a private school in Paraguay that cost $200 a month. Though she yearned to hold Guido in her arms, instead she paid for his swimming lessons, the earmark of the privileged. The money she sent home helped her mother pay for concerts and go to the pool with her son.

Activist movements are raising public awareness of domestic labor, but women like Blanca are still overlooked, despite the battle for better wages and despite the fact that they are the major players in one of the fastest-growing sectors of the U.S. economy. There are millions of caregivers in America, but because of their immigration status, they may live discreetly under the radar. According to the Bureau of Labor Statistics, 1,260,600 child-care workers were employed in the United States in 2011. However, 8.2 million children under five spent time with a child-care provider other than a parent, such as friends, family members, or Head Start programs. Obviously, then, a great many child-care workers out there are unreported. This rapidly growing sector has pushed to the forefront the situation that Blanca had to grapple with every day: if more and more caregivers are needed in America, what will happen when so many families—so many parents and children—are living so far apart?

Such family separation had not been Blanca's experience back home in Paraguay, where her family and so many others had intimately interwoven lives. That closeness back home was forged, however, out of hardship. Blanca had grown up very poor: she received Christmas presents for the first time when she was nine, and for years she made her own dolls out of corn husks because no one could afford to buy her one.

At night, her mother would tell her and her brother to lie in their beds while she found them food. It was the trick of the mother with nothing—while Blanca and her brother waited for food that didn't exist, they would fall asleep. At age eight, she began to help her mother do everything at work, cleaning and cooking for another family. As an adult, she fell in love and gave birth to Guido. When she discovered that Guido's father had cheated on her, she left him.

A single mother, Blanca realized that she had to support her son and her mother on her hospital nurse's salary. (Guido's father was an inconstant presence in his son's life, as she described it.) It was impossible. According to Blanca, nurses did everything at the public hospitals, sometimes without hot water or electric light, for very little pay. Struggling to make a living, Blanca made her way to a nannying job in Miami, with employers who gave her no vacation and paid her extremely low wages. She was a "live-in," in the parlance of domestic labor. "I am working like an animal, three children in Miami, cooking, ironing, cleaning, and organizing," said Blanca. She then moved to New York City, where she found a $15-an-hour nannying job, gained her citizenship, and added a job cleaning houses on weekends for extra cash.

In the winter of 2014, I met with a number of care workers like Blanca in New York City. There was Esther Simiyu, who grew up in Kenya; later that night she would head to Manhattan's Upper East Side to stay up all evening with a newborn while his mother slept. As of 2014, Esther's husband was in Nairobi with their fifteen-year-old daughter and eleven-year-old son. The first time she left them had been nine years earlier, when she came to the United States on a student visa; she spent the initial few months crying and sleeping fitfully. Esther made enough money as a baby nurse and sleep trainer—$20 to $25 an hour—to send her kids to a boarding school in Kenya. "Now my daughter is a teenager. We are Skyping, and she says, 'I want to ask you a question, Mama,' and I want to hug her, but I cannot." The separation is painful for mother and daughter. "But it is important that I am doing this," said Esther, slapping the table. Her wages were supporting eight family members, including her two children. She sent home checks in varying amounts, sometimes for as much as $1,000. She also shopped for gifts for them in Manhattan's Times Square or the Bronx's Fordham Road. Her husband worked, she said, but somehow, as with so many of these accounts, the pressure appeared to fall on her, the mother.

For these newcomer caregivers, keeping their children with them in the city poses severe financial challenges. According to a 2012 study, 70 percent of the domestic workers surveyed were paid less than $13 an hour. And even for those who are paid adequately, the cost of child care can make keeping their children near them impossible.

Pema, a nanny from Tibet who worked in Manhattan, was sending her two-year-old son back home, as of April 2014, because she couldn't pay for his day care in New York City. As a working nanny, Pema said, she often needed to be on the job at 8:00 A.M., while the day-care drop-off time for her own son was also 8:00 A.M.

"Timing is so hard," Pema told me. Her dilemma stemmed from another semiconscious expression of class privilege: buyers have the power to dictate the family style of those they hire and often view a nanny's children as a potential hindrance. The caregiver with a child is less flexible than the childless care worker. This disadvantage in the nanny market runs in nasty parallel to the corporate impatience with pregnant women that we learned about in the first chapter. "When you say you have kids, most moms who might hire me as a nanny prefer to hire nannies without kids," Pema said.

BLANCA HAD TRIED BRINGING HER SON TO NEW YORK WHEN HE WAS nine, but Guido had hated living in the city so much that he ultimately went back home. Blanca was traveling a lot for work, and he'd had to stay with a friend when she was away. Guido did not like school either, and the other kids made fun of him. He missed his grandmother; after all, she was the one who had been raising him. Given her working hours, Blanca was unable to spend the time with her son that they both wanted. When Guido returned to Paraguay, he and Blanca developed a pattern of frequent phone calls, sometimes talking ten times a day.

When I met Blanca in late December 2013, more than two months before Guido's arrival, she was looking after the verbal nineteen-month-old she nannied and clearly adored. The toddler sat in his UPPAbaby stroller, eating Goldfish crackers out of an orange container and playing with a Popsicle stick; every so often he'd say, "I am hot," or, "Water," and drink from his sippy cup.

Blanca loved this boy. She oohed and aahed over his early speech, sang "You Are My Sunshine" to him, and proudly showed pictures of him playing in the snow to people who expressed the mildest interest.

"It's hard to make money, and I always take care of a baby," Blanca told me. "Maybe because I don't have my son, I take care of children, because I miss him. I have missed his childhood."

Thirty years ago, it was mostly Caribbean women who worked as caregivers; their own children were dubbed "barrel children" because they received giant containers full of gifts and clothing from their transnational moms. Tamara Mose Brown, a sociologist and the author of *Raising Brooklyn,* a study of caregivers, said that today caregivers are increasingly from Latin America or places like the Philippines. When I met Brown in a Brooklyn Tibetan café, she explained that caregivers sometimes send 60 percent of their income to their families; these remittances not only provide important support for their families but also stabilize their home countrys' economies. "The money provides tuition and better food choices for the children as well as material goods," Brown noted.

But what happens when the families yearn to be together? Family members seeking to reunify with their children, spouses, or parents make up the majority of immigrant petitions filed and visas granted each year. The children of people who have become American citizens are granted visas without regard to quotas. In theory, it should be possible to approve a visa petition for these children quickly. In reality, however, huge backlogs have created lengthy processing delays. This has only gotten worse with the anti-immigration stance of the Trump administration, which has deported even overseas adoptees who thought they were American citizens, usually thanks to faulty paperwork; this shocking practice is likely to become even more common if Donald Trump stays in power. This predicament exposes a major contradiction of globalism, that is, the way in which labor is interconnected throughout the world. The attitude of some of the anti-immigrant affluent that seemed ever clearer in the Trump era is that they may be happy to have migrant workers doing their chores and minding their children but it simply *won't do* to have laborers' families coming to the United States to live.

Once here, these workers also may lack the public or familial support scaffolding that would help them care for their own children. Back home, families tend to take over where the state falters. As Esther, the baby nurse from Kenya, put it: "I come from a collectivistic culture; you guys [Americans] are individualistic. I pay for my aunt, the colleges of my whole family." These family members, in turn, keep tabs on her children in her absence.

Blanca's and Esther's struggles bothered me, for personal reasons. I am the granddaughter of striving Polish and Russian immigrants, and my grandparents on both sides came to America when they were young and unable to speak English. Wrenched from their families, some as young as teenagers, they believed—correctly—that their hard work would lead to social class ascension, that they would become middle-class Americans, and certainly that their children would fulfill their highest hopes. Such aspirations are no longer fulfilled. It was tough for my grandparents, and they encountered some barriers that no longer exist, or not in the same form—such as my grandmother's inability to find work as a teacher, despite learning English perfectly, because she spoke with a Polish accent. Still, my grandparents seemed to believe entirely that opportunity was there for them. Today people like Blanca are not so sure. What changed?

Blanca and the many women like her who have left their own children overseas in order to work by taking care of American middle-class children—whose parents are sometimes squeezed themselves—show us a new American parenthood. The middle classes, at least in the last 150 years, have constantly been refreshed and invigorated by infusions of new people, either the working class or immigrants. But many indicators suggest that *los de abajo* (underdogs) like Blanca are having a harder and harder time finding routes of entry. If the American middle class remains seldom, if ever, refreshed, it won't survive.

Blanca had been a nurse in Paraguay, with an advanced degree, but that credential didn't carry over to the United States. There were holes in the social class chain on which Blanca was a link, and immigrants like her often fall through these gaps. Blanca would have had to do something like become a nurse in order to enter the American middle class, but she couldn't do that without the necessary training to get certified. Such training remained out of reach, however, because she couldn't afford it, she lacked the time to go to school, and her English wasn't strong enough. So she remained a subsistence laborer.

Blanca's situation demonstrates one of the problems immigrants face today that they didn't in the 1950s—the likelihood that attending a higher education institution of any kind will incur significant debt. By contrast, my grandparents had access to abundant and either free or very reasonably priced college education in the earlier part of the twentieth century at places like City College, where one of my grandfathers went to night school. City College was then known as "the proletarian Harvard." Immigrant Jewish students went there at little cost and as an alternative to the Protestant elite universities that rejected them.

The middle class has always been preserved and created, of course, through broken links: many immigrant populations came from their home countries on their own, at young ages, leaving behind parents and siblings who were hungry, unemployed, or even being harassed and killed. They had to fill in the familial gaps that their migration made in their own new families on American soil with the care of extended family members or through social networks in their neighborhoods. But today the people most likely to help working-class immigrants like Blanca who aspire to the middle class are less likely to be able to get into the United States and, if they do, more likely to be unable to stay.

Being an immigrant hasn't always kept newcomer parents in America down or stalled. My immigrant grandparents had a small shoe store in the Bronx, and as a child, I played on the floor of the store with polish, a shoehorn, and a footwear stretcher. As a teenager, my paternal grandmother worked on New York City's Lower East Side putting feathers on hats. In the early 2000s in the East Village, the primary music was a steady buzz of receipts being printed out in thousands of franchises and boutiques. But in my mind, figures of those early-twentieth-century workrooms hung around the neighborhood like ghosts. I thought of how my mother got through college by working at a department store, and then how I got to be a special flower in progressive school thanks to her determination that my life would be full of poetry recitals and those colorful math counting blocks they called Cuisenaire rods. These are transformations of first generation immigrants that are harder now.

Immigration matters in the question of inequality and how American families get squeezed. In fact, being an immigrant often caps your mobility, and being undocumented—or even a new American citizen—can be far worse. It wasn't always thus. As Kessler-Harris told me, "I am an immigrant: I benefited from the 1950s expansion of education, the tremendous opportunity for jobs then. The public school system had a lot more credentials and faith. If you got an education, you could move up in the world. That was realistic. And in 1910, if you came to be a nanny, maybe you found a male partner and moved up in the world, and moved with him: that is no longer possible, not even with a two-income, two-partner family."

Kessler-Harris enumerated why so many people in this country no longer live the American Dream, how that differs from immigrant experiences of the past, and why one more path to middle-class status is no longer available. One element is the difficulty of immigrating: in 2017, a cruel year for immigration to America, over 1 million family-based immigration petitions were pending; some had been stalled for years. (As the Trump years progress, this number will surely only become larger.) Moreover, not every employer acknowledges the harsh obstacles their employees face. "My boss is very nice now—she asks, 'How's Guido, how's your week?'" Blanca said when I spoke with her before Trump began campaigning for president. "Some people don't care if Guido is still alive or not." Since his election, her words have returned to me, now literally come true.

New efforts have been made to raise the consciousness of employers of workers like Blanca. The organization Hand in Hand tries to improve conditions for domestic workers by laying out standards for pay and time off and encouraging employers to understand the life of the caregiver who enters their home; there should be no delays in payment, for example, because their employees send money home to their kids, who are depending on them for money to buy schoolbooks. Besides these guidelines for employers, Hand in Hand also encourages caregivers to enjoy small comforts while on the job, like cooking their own lunches in their employer's kitchen. One of the group's advocates, Gayle Kirshenbaum, told me that she started out as a "confused employer of a nanny—I didn't have the resources to be the employer I wanted to be." She added that, for care workers, their "home is also a workplace."

WHEN I MET BLANCA AGAIN IN JANUARY 2014, SHE THOUGHT THAT things were about to change for her and her son. She wanted to bring him back in March, whatever the consequences, because her mother was just too old to care for him any longer. "He's not going to have it easy in America," Blanca said. While reunification sounds good for families, it presents its own difficulties. Guido spoke very little English, and Blanca's hours didn't permit her to be home before 7:15 P.M. Could she work fewer hours? She worked on weekends, after all, doing additional child care and cleaning. What if Guido got sick? "I need to be able to stay home with him."

Blanca called me again in February 2014. During a momentary dip in the price of air travel, she bought a one-way ticket for Guido. He would be arriving in three days, accompanied by Blanca's best friend Gloria—who, although surviving on a nanny's salary herself, had paid for her own round-trip ticket to Paraguay as an act of friendship. Blanca could not afford another ticket for herself, and she also couldn't afford to miss the number of days of work that the long round-trip flight would have required.

With Guido's impending arrival, Blanca's work was cut out for her. His future room was cold and windowless, and its floors were still unfinished; her friend would be helping her fix them the day we spoke. She prepared for Guido's arrival by cleaning her house—sweeping the floors and washing them with a mop that she created from a stick and rag, doing for her own home what she had done for many other people's homes that year. She gestured at a long leafy ivy plant in a china pot that stood in one of the only two small windows of her apartment. "This plant has been my company since I moved here—that plant is my best friend," she laughed.

Then there was the fact that she hadn't seen Guido for two years—buying the new clothes he'd need would be difficult. "He's taller than me," Blanca said. Guido didn't Skype with her, since he didn't have Internet access, so she had trouble picturing him sometimes. She had bought him boxers, socks, and pajamas with hearts on them, but she didn't know his size, and "he's very boy, so I am not sure he will like these hearts." At some point, she called Guido as he prepared for his journey and talked to him in Spanish about the clothes he was packing. He told her that he was packing summer clothes and she laughed. It was snowing again in New York City.

"I feel bad for my mom," said Blanca. "I know it's for the best that he leaves, but sometimes I feel like everything is neither good nor bad. I hurt my mother. I hurt Guido. And now: what school do I take him to? I have no ideas. I ask at the school nearby, but no one speaks Spanish. I don't know what is safe. His life is so much safer in Paraguay."

Guido had not lived with her for more than a few months at a stretch since he was about two years old. Blanca had been thousands of miles away from the city where he grew up. She was in a cold city that had more people in it than his entire country. "I only think of you: I want to see you again," Guido told his mother by phone.

ON THE DAY BLANCA WENT TO MEET HER SON AT THE AIRPORT FOR their reunion, she was all nerves. In the long drive out to Queens, and during the hours spent in the airport lounge, Blanca told me stories of her old life working as a pediatric emergency nurse in Paraguay, where the hospital lacked basic equipment. She told me about the time the other nurses and doctors thought a six-month-old was dead and she saved the child's life with oxygen. I thought of how she had gone from skilled work to the more informal work of nannying—it was a common trajectory for immigrant women's labor, as the scholar Karen Brodkin and the National Domestic Workers Alliance have noted.

Blanca's memories of these times both sad and happy in the past were repeatedly interrupted as she checked for messages on Guido's arrival.

"Lately, I lose keys, lost my Facebook password, and forgot my friend's birthday," she said. "Guido, my mother—no one knows how much I worry."

The long minutes stretched into hours. As each person went through immigration, Blanca jumped. Two hours after the plane was supposed to land, she saw her son and Gloria. Her face lit up. "Hey, Guido! Guido!" she called. Although the plane arrived on time, immigration agents had delayed the pair. They kept asking them how Guido could be an American citizen when he didn't speak English.

"*¡Mi amorcito lindo!*" Blanca said, kissing Guido. Her son looked dutiful but happy. He didn't talk much. He was here, Guido said, "*estar con mi mamá*."

Blanca took the dark blue jacket she had bought for him out of the bag. It wasn't his favorite color, red, and he didn't really want to wear it, he said. But it was freezing outside, she told him. Guido was carrying just one suitcase, containing two pairs of jeans and one pair of sneakers.

"Put on the jacket, Guido. Poor boy, they treated him badly in immigration," Blanca said in Spanish.

"*No tengo frío*," Guido said: he was not cold. Also, the new coat didn't fit.

"I can't believe he's so big!" Blanca said. "I have to return the jacket."

They all needed a break from the cold, but the cab's warmth would last only a little while. They were going to Guido's new home.

By March, Blanca had gotten Guido enrolled in Leonardo da Vinci Intermediate School, a new, 2,270-student middle school only a few blocks from Blanca's apartment, in the working-class neighborhood of Corona in northern Queens. The largest middle school in any of the five boroughs, Leonardo da Vinci was eligible for Title 1 federal funding because of the student body's high poverty level. Its teachers rolled trolleys overloaded with schoolbooks while students filled the corridors, heading in all directions. Pairing upper-class aspirations with a low-income student body, sections of the school were named Princeton, Stanford, Yale, and Harvard, the very embodiments of educational luxury as well as harbingers of excellence to aim for.

Blanca and others like her have helped revive neighborhoods like the one in which Guido went to school. But rents have also risen, and now immigrants tend to live far from the city centers where they tend to work. Technology, banking, and investment firms now operate in centralized locations, but as Nancy Kwak, a historian at the University of California, San Diego, and the author of *A World of Homeowners,* explained to me, these industries have attracted a flow of affluent people and businesses into key urban locations, raising land values. Yet those who work in the lower tiers of the same industries—in hospitality, or caregiving—have the same demand for housing in these areas, but are unable to pay for it.

The low valuation of caregiving as a job was one reason why all of Blanca's domestic work had done little to secure her future. The vacuuming, the warm bottles offered to children, the loving or punishing gazes dispensed—all these aspects of caregiving work are routinely devalued and usually underpaid or unpaid. For centuries and across the world, women's domestic labor has usually not been compensated, there being a deep-seated belief that women "naturally" serve others gratis. As the scholar Paula England has written, a common perspective on caregivers argues "that care work is badly rewarded because care is associated with women, and often women of color." As England notes, "The devaluation framework emphasizes that cultural biases limit both wages and state support for care work. Today mothers still do most of our unpaid work." (A 2015 McKinsey report estimated "that unpaid work being undertaken by women today amounts to as much as $10 trillion of output per year, roughly equivalent to 13 percent of global GDP.")

American parents, it is estimated, do chores and child care for 1,183 hours per year, but half of American women do housework each day, while only 20 percent of men do; annually, working mothers put in roughly ten more days of household drudgery than their male partners, and twice the amount of child care. And working mothers who are among the huge number of single mothers do it all. According to a 2013 Pew Research Center study, one in seven Americans care for their children and their parents simultaneously. Mothers do fifteen more days of care work per year than fathers, according to the Pew study (and we're talking twenty-four-hour days).

The BLS American Time Use Survey found that the time women spend cooking and taking care of their homes hasn't changed much since 2003. Women spent about 2.24 hours on these tasks each day in 2016.

The history of maternity and women's care work is a story written in statistics as well as in nappy pins and diaper rash creams. But there have also been more creative reactions to mothers' oppression. The American artist Mierle Laderman Ukeles wrote in a 1969 manifesto: "The culture confers lousy status on maintenance jobs = minimum wages, housewives = no pay/clean your desk, wash the dishes, clean the floor, wash your clothes, wash your toes, change the baby's diaper, finish the report, correct the typos, mend the fence, keep the customer happy." Ukeles turned her own exhaustion and frustration at the devaluing of her labor and the labor of those around her into poetry and performance art.

In the 1970s, Wages for Housework amplified Ukeles's aesthetic impulse into an unsparing movement. Wages for

Housework argued that caring for one's own children and cleaning one's own home should be compensated and politically and emotionally recognized. Italian-born Silvia Federici was one of the movement's central figures. Federici, now a mild-mannered professor of seventy-three who lives in Park Slope, spoke with me in a hypnotic lilt, but she remained obsessed with rescuing women from drudgery and also with the ongoing undervaluation of domestic work for hire. In my conversations with her, she noted the unpaid work that women do: as they cook and clean, smiling and often bending over backward to please in social situations, in the workplace, and in their intimate partnerships, women ultimately lose some of their earning power. She murmured that women's domestic and emotional labor in the home could be considered a form of domestic violence.

This sounded extreme. But don't women generally deserve some greater reward for the invisible domestic, caring, and emotional labor we do, either that which is poorly paid or underpaid? I was so pleased to hear a radical counterpoint to concepts like "leaning in" and all the other mainstream memes about female corporate empowerment. "These ideas are so superficial that I haven't given them much thought," said Federici, with something of a sneer. "'Having it all' is an ideologically distorted conception. What women are missing are autonomy and money of their own. They are also completely overworked."

AT FIRST, THE NEW LIFE IN QUEENS WAS DIFFICULT FOR GUIDO. HE missed his grandmother, now ninety-two. They still spoke every day, and he kept photographs of her on his phone. In his favorite one, she was seated in a housedress, looking lovingly into the camera.

He sometimes despaired, Blanca said. Every day after school, he would walk home to an empty apartment in a building with crowded rentals full of immigrants cooking and talking on every floor. There he'd watch the hours tick by, listening to the din through the thin walls, until his mother got home.

Things gradually got better. Blanca bought Guido a puppy, an expensive purebred shih tzu whose very cost and fastidiousness seemed to represent *success in America* to Blanca. Guido began to make friends. He grew tall and good-looking, with a thick wave of hair. He was also good at soccer, an invaluable skill that could turn a kid into a monarch in a school of immigrant students. One day when I visited him at school, I watched Guido's friends gather around him as he kicked the ball and blocked his opponents. He was wearing chinos, new red Nikes, and a polo shirt, and he towered over the other kids. Shone, the cool teacher who supervised recess (with a buzzed-on-the-sides hairdo), told me that the girls liked Guido best. His friends Dilan from Colombia and Juan from the Dominican Republic clearly looked to him for his judgment on social matters.

As the eighth-grade year progressed, the chatter among the students began to focus on high school. What would it be like? Which ones would people be going to? In one of Guido's classes, a girl with turquoise nails, styled with meticulous designs that she found on the Internet, leaned over to a friend who was putting on lip gloss. "Forest Hills High School. Flushing High School," they murmured. "That's what my mother found out. That's what my cousin said." These were more suburban and highly thought-of schools than the school where Guido was likely to end up. The girls occasionally glanced at Guido, sitting refined and contained at his desk.

In his math class the same day, Guido took careful notes. Occasionally, he translated the teacher's words into Spanish for his more-recent-immigrant friends. Guido's energetic math teacher, Samantha Heuer, who sported leopard-print boots and stylish ombré hair, taught linear equations. The keys, Heuer said in a bright and commanding voice, were indeterminate and determinate variables. "In the equation involving the boy Juan," she said, "the Y variable is indeterminate and X is determinate."

Heuer asked Guido to solve a problem for the class at the board, and he did it with ease. Guido was a cherished rarity for Heuer: a student who would come to her after class and ask how he could improve his grade. Guido did best at math. Yet for all his efforts, his grades were mediocre, topping out at low Bs, and he had just received a 54 on a science test.

Guido's life itself resembled a linear equation, with determinate factors and indeterminate ones, Y and X variables, like in math class. The good determinate factors were his inherent physical grace and the love of his mother and grandmother. The bad determinate factors were the language barrier and living in near poverty in a fractured family. And the indeterminate factor was luck.

When working-poor New Yorkers like Blanca and Guido seek spots at the city's more desirable schools, they often must compete with middle-class New Yorkers. The competition isn't set up to be even.

So many challenges face those who are knocking on the door. There is a persistent failure in urban schools to find bilingual communicators, even in the most diverse cities. One definition of being middle-class is not only being adequately literate in English but knowing how to locate and access services in the language. America's social service sector doesn't tend to share how it does things with those who don't already somehow know. So those who don't already understand how education, health care, and child care are administered and how to negotiate these systems may lack the access that a middle-class person has without even working for it. It becomes a big game of "If you don't know, I'm not going to tell you." This brings us back once more to cultural capital—the knowledge necessary to figure out unwieldy social systems using one's networks.

THE MOST PRIVILEGED PARENTS CAN HIRE TUTORS AND SEND THEIR children to expensive prep courses. Others hire coaches who have built careers out of helping parents navigate public school choice. I once saw a mother buttonhole a principal and insist that he rank all the public schools in the mother's catchment area in descending order. As she took up the man's time and ignored the line of other parents waiting to speak to him, a phrase popped into my mind, a variation on the "will to power" of philosophy: the "will to education."

Most of the parents I know—almost all of whom, admittedly, can be categorized as upper-middle-class—spend a lot of time decoding school statistics. They look at everything from test scores to absenteeism, searching for indications of a school's strength. I have tried to read between the lines to discern whether one school or another would be good for my daughter. I have also scoured school and parent websites; the latter have sometimes startled me when parents itemize local public school failings for one another or half boast about what their seventh-graders are doing for "extras" to add to their appeal on their New York City public high school applications. These "extras" include activities like "cell replication" and learning Mandarin. All of this takes a lot of social and systemic literacy as well as time.

Parents in the privileged classes are far more likely to know their way around, to know what they need to know, and how to find out the answers. They are much more likely, just by being who they are, to know the silken, tasseled ropes of the choicest schools.

I was so curious about how the middle class handles the obstacles that Blanca faced that I tracked down Joyce Szuflita, one of New York City's most famous public school consultants, to discuss the contrast. One summer day in 2015, I asked her: how do those who have the capital to navigate the schools do it?

Szuflita was a cheery, graying blond woman of fifty-six who wore a necklace of heavy, brightly colored beads. As a public school consultant, she provided soothing information and swift judgments and directed parents through the city's byzantine public system with precision. There were so many questions at sessions like these: Zoned or lottery admissions? Regular public or charter? Which schools have families like ours? Which ones have room for kids who live in another zone? What's most effective in getting a child off a waiting list and into a seat? The parents I observed working one-on-one with her, as well as those I watched during one of her frequent public speeches, hung on her answers as if she were a cult leader. At one diner meeting in Brooklyn, a client couple—an academic and a public-health researcher—had come to her with a dilemma. They were moving from Cambridge, Massachusetts, to New York City and were nervous about the schooling of their two-year-old and their not-yet-born child. They knew that securing a spot in one of the borough's handful of coveted high-performing elementary schools was notoriously difficult. Szuflita's clientele consisted mostly of college-educated professionals like this couple—parents who were trying to balance their hopes for their children with the challenges of raising a family in one of the country's most expensive cities. (She also consulted on private schools.) A meeting like the one I observed in a diner over iced coffee, hash browns, and a small pile of district maps can cost $400 to $550. (At the budget end, the fee for fifteen minutes on the phone with her is $50.)

That a consultant like Szuflita is so sought after indicates the broken links in how school systems communicate with students and parents, as well as the expectations of privileged parents. That consultants for public schools are kept busy is a testament to a bewildering system that is increasingly divided by race and class and harbors both obvious and hidden inequalities.

"This school would be more of a safety," Szuflita explained to the two parents meeting with her. She gestured at the colored paper handouts fanned out between them, then pointed at another zoned school on a different sheet. "That's a curated class of parents because they chose to move into the zone of the school that they wanted," said Szuflita. "So you don't have to have G&T if you have that," she added, referring to gifted-and-talented programs; in other words, she felt that a regular school with an affluent and committed parental body could easily be as good as a G&T school in New York.

Public school counselors like Szuflita tend to serve upper-middle-class parents in large cities like Los Angeles, New York, Chicago, and San Francisco. I saw how valuable the information that Szuflita offered parents was.

But what if such a useful form of knowledge, cultural capital, and support was available, at a far lower fee, to people who are financially unsteady? What if both working-class and impoverished parents who know little (and are told even less) about test dates, application days, and the relative benefits of various public schools could turn to consultants like Szuflita, on the Department of Education's dime? What would students' outcomes be if school consultants like Szuflita were available and widely and fully subsidized for less well-off families? And most importantly, what would an adequately funded public education system look like, not one with truly bad and truly good schools? That kind of scholastic and class fairness would render Szuflita and the like obsolete.

Kindergarten admissions in places like New York City are mostly decided by zoning, plus a combination of lotteries and choice. Popular schools can be difficult to get into: class sizes have gone up, and some schools have waiting lists. Gifted-and-talented and specialized programs, such as dual-language classes, admit a small number of applicants. Finally, the difference between a "good" public school, where a PTA raises up to $1 million a year, and a "normal" public school can be the presence of basics such as substitute teachers. In prosperous public schools, what were once basics are now sold as if they are value added.

People like public school consultants know things that overwhelmed parents would never have time to deduce. A relatively easy correction to achieve a more fair school playing field would be offering parents help decoding the public school system, either for free or at a highly subsidized rate. After all, many elementary schools don't feed into the nearby middle schools. Parents might be unaware of school zone changes or simply not know the significance of a school's attendance rates: for instance, high absenteeism—around or above 30 percent—suggests a floundering student body.

On the surface, public schools can seem egalitarian, especially when their websites emphasize words such as "connection," "community," and "choice." Yet despite this democratic vocabulary, money makes a big difference. The list of costly services that supplement some children's public educations—additional art lessons, fencing lessons, music classes, and even math and test tutoring—is growing longer and now includes consultants and test prep. That's in addition to the homework help some stay-at-home parents can provide.

In New York City, the entire system is particularly racially and class segregated, even more than the city itself: by some statistics, New York has the most segregated school system in the United States. A recent report out of the University of California, Los Angeles, found that in nineteen of New York City's thirty-two public school districts in 2010, 10 percent of the students or less were white, which means that the majority of Caucasian parents opted out of the system altogether.

Sometimes a school consultant can ease parents' worries by reminding them of their privilege. "As many as twenty-eight thousand kids will be taking the test for schools, but most have never done prep," Szuflita told a crowd of parents who, despite a snowstorm, had gathered in Brooklyn to hear about high school admissions, many wearing heavy knit sweaters and fashionably worn jeans. "They think they are smart, but they walk out halfway through. They are not all your kids, who are beautifully prepped!"

"My work is about keeping the middle class and upper-middle class in the public schools," Szuflita told me on the phone in October, her "crazy" school application month. She was implying that by at least allaying concerns about admissions, she and others like her were helping retain families who might otherwise leave the public school system. Studies have found that there are benefits to public schools when they are more economically diverse, and in a city where the top 5 percent earn close to $900,000 a year—or eighty-eight times the median household income of the bottom 20 percent—keeping some wealthier families in public schools could be considered a political act. On the other hand, and not unexpectedly, the schools where wealthier families end up tend to be the least class and racially varied, and the most buoyed by parental giving.

Szuflita acknowledged the problem. "I am consulting with families who can afford me," she said. "I am well aware that there are people who can't [afford to] use my service, but I am also trying to put food on the table of my family. So I serve families like me." Daniel Janzen, a client of Szuflita's, spoke of her glowingly, but mentioned a similar concern: "We are looking for premier schools, but why isn't every public school great or a top choice?"

Szuflita's world was far from Blanca's. One night in late 2015, the streets of their neighborhood were dimly lit; the brightest lights came from barbershops and bakeries displaying giant, garishly colored birthday cakes. The residential blocks were mostly multifamily houses, some new, many built in the 1920s. Guido and Blanca's home was between an NSA supermarket and a parking lot. Guido was gearing up to write out the list of high schools he'd apply for.

Blanca didn't return from work until the evening. Nonetheless, she checked in regularly, calling Guido throughout her train ride home, apprising him of her whereabouts and checking to see if he was WhatsApp-ing or playing video games like *Call of Duty* instead of doing his homework. When Blanca made it home, she gestured around her neat, dark apartment in a tour-guide pantomime. "Welcome to my penthouse," she said to me.

Within ten minutes, she was sitting at the small kitchen

table and going over Guido's homework. She looked at the math problem sheet, less to judge its quality (Blanca hasn't studied math in years) than to ensure that it was finished. How much time had he spent on it, she wanted to know. Blanca muttered to me irritably that Guido devoted only an hour to all of his homework. She had been educated in a very different country and era, and she was not sure what to tell Guido except to follow her example of hard work.

Recent studies have questioned whether there is any actual evidence that mobility lessened after the post-1979 rise in inequality. Part of what happened then was that the income of the top 1 percent increased astronomically: between 1979 and 2007, earnings for the most privileged soared nearly ten times as fast as those for the bottom 90 percent. If American incomes now looked like they did in 1979, families in the bottom 80 percent of the income distribution would be making an average of $11,000 more per year, as calculated by Larry Summers in the *Financial Times* in 2015 for a piece optimistically entitled "It Can Be Morning Again for the World's Middle Class." If it were 1979, the top 1 percent would make $750,000 less. In other words, today the class rungs are spaced much further apart, which makes the ladder harder to climb.

Blanca heated the pizzas she had brought home—thin-crust gourmet leftovers from her employer—slice by slice in the toaster oven, which stood in for a full-sized range. Guido had shed his immaculate preppy uniform, mandatory at Leonardo da Vinci, for an Angry Birds T-shirt, and he was texting up a storm.

After dinner, Blanca reached for the *Directory of NYC Public High Schools*—"the book," as she called it, almost reverently. She and Guido looked briefly at the giant tome, then closed it. They seemed to have trouble making sense of it. On the state Education Department choice form that was due in a few weeks, Guido had simply written down the schools that sounded best to him, based on his interest in business. In New York, different schools cater to teens with different interests, but like a novice bettor picking horses, he picked schools strictly on the music of their names. His first choice was the High School of Economics and Finance. Further down his list was the High School for Arts and Business. What about LaGuardia, Flushing, and Forest Hills, I asked, although I later realized that Guido might not even be zoned for them. I turned to Blanca. She told me that she had no idea which schools to push for.

How were immigrants like Blanca to understand the hidden and not-so-hidden meanings of the school websites? Blanca and other parents who don't read English fluently are hobbled by not having enough translated materials about schools. Blanca was also uncomfortable speaking with school personnel. I remembered a night a year or more before when we were at an Italian restaurant celebrating Guido's arrival. That night she had told me how nervous she was about his schooling. By this time, Guido's English was a little better than Blanca's.

Blanca was also hampered by how little work flexibility she had. She couldn't take time off from work to meet with teachers, school counselors, or officials, to visit potential schools, or to lobby for Guido's admission. Blanca also didn't know about the many educational services available to her, which could have helped with school placement or tutoring.

In March 2015, I got a text update from Blanca. She wasn't happy. As I soon learned, Guido had been admitted into his first choice, the High School of Economics and Finance. But Blanca hadn't realized that the school was in downtown Manhattan, nearly nine miles from their house, and Blanca worked seven miles away from the school in a different direction. While Guido could take public transportation to the school, his commute would take close to an hour. She worried that he might not yet be ready to do such a long daily commute alone. Blanca was most likely wise to say no to a school that was far away from her, even though many parents are even farther away from their children during school hours. Her voice rose, as it often did when she was anxious.

There was a fair for kids who weren't happy with their school assignment and were hoping to be reassigned, I told Blanca. Shouldn't they go?

On a pigeon-gray rainy day in mid-March, Guido, Blanca, and I met up at a Starbucks to make our way to the Round 2 High School Fair in Manhattan. "I was so stupid to let Guido apply to that school first," Blanca told me. "So stupid!" I bought her a latte, a favorite of hers. It was her forty-fifth birthday. Soon we were walking through Lincoln Center. Guido had never seen the grand theaters and plaza. He and Blanca peeked at the giant chandeliers and well-heeled opera aficionados in the stately theater buildings.

When we arrived at Martin Luther King Jr. High School on Sixty-Fifth Street and Amsterdam Avenue, we discovered that Blanca and Guido were far from alone in their quest. Long lines of kids and their parents snaked around the block, waiting to get into the fair in the hopes of correcting a school assignment that had been found wanting. After standing in the rain for twenty minutes, listening to conversational bursts of Spanish, Russian, French, Mandarin, and English, the three of us finally got inside. Colored arrows directed students to schools in their boroughs, and with the help of a translator, Guido and Blanca found the floor for Queens. Throngs of students and parents circled around tables manned by school representatives.

A young man directed us to another floor to get help from a school counselor. The counselors were seated in an auditorium with red, orange, and purple balloons and plastic tablecloths in matching colors—the setting resembled a prom. The overall effect was disconcertingly festive given how distressed Blanca and Guido were. One counselor said that she spoke Spanish, but she labored to understand as Blanca, frustrated with not being fully understood in English, fell into her native tongue. Though she spoke English, when Blanca was trying to express nuances it sometimes failed her. Her face flushed. She felt faint, she said, and needed water. She feared that Guido's education was doomed.

Could Blanca appeal the school decision? The counselor told us that she could, but that the deadline to do so was fast approaching. It was the first time Blanca had heard of such a process. Then the counselor told us about a high school called Newcomers, open only to Queens students or residents who had lived in the United States for one year or less at the time of admission to high school. There would be a soccer team for Guido and a media center, and it was pretty close to their apartment. "This could be okay?" Blanca said, her statement, like so many involving schools, phrased as an open question.

Then the counselor warned us that the Newcomers representative might not be at the fair. "It's a school for new Americans," she said, "and new Americans have often not heard about school fairs."

We were directed back through the throngs of students and parents to the floor for schools in Queens. The counselor was right: no one from Newcomers seemed to be on hand. We found one school in Long Island City that might have openings, but Blanca was nervous about its quality. And while she was given a date to attend a tour of the school, Blanca knew she wouldn't be able to leave work early.

Blanca and Guido's circumstances are in many ways quite ordinary. As of 2015, more than 37 percent of New Yorkers were foreign-born, and many lived on lean budgets. But the ordinariness of Blanca's and Guido's lives in New York also conceals what can feel so extraordinary for all these newcomers. "By any measure, immigration is one of the most stressful events a family can undergo," writes Carola Suárez-Orozco and Marcelo Suárez-Orozco in their book *Children of Immigration*.

Guido loved being in the United States, but Blanca remained worried. After more than a decade here, she had grown familiar with (and sometimes partaken of) certain tangible luxuries of American upper-middle-class life, regardless of whether she could afford it, such as good coffee and crepes, Whole Foods, and the $1,200 purebred pup she bought for Guido (roughly two times her weekly salary at that time). Some of the intangibles of middle-class life, however, like the savvy required to get into an optimal educational track, still eluded her.

Suárez-Orozco thinks that the first step in incorporating immigrant children into public schools is to make the school choice process easier to understand. It is common for immigrants to arrive in the United States "thinking American schools are the best in the world, to come with basic trust, and assumptions about educational authority that are not fully true," she said when we spoke. Schools need "to create more transparent educational pathways," she added, "so you don't have to be a chess master to get your kid through school." It's not only immigrants, however, who must navigate the school choice process. The experience of parents and students like Blanca and Guido should match that of the privileged families who wend their ways through the public school system, knowing how to lobby for access to, say, the progressive public school with the farm on the roof, or the "best" (and wealthiest) public schools whose zone borders encompass only the most expensive addresses.

These educational barriers are different from the ones faced by earlier generations of immigrants to the United States—today's barriers stem from overall inequality, and the invisible barriers like cultural capital are often the problem. School consultants and the like are important because their services are integral to how the relatively privileged maintain their "good" places in the public sphere. School consultants themselves, after all, are a form of cultural capital.

In the early 1900s, public schools began to offer office training for girls, and after the passage of a vocational training act in 1917, secretarial training in public schools expanded rapidly. Nurses' training was often offered through apprenticeships and hospital-based programs; students labored in hospitals while they studied to become nurses and sometimes paid no fee for their training. These programs made mobility into the middle class somewhat easier than it is today. This is another part of the squeezed story: all the barriers to entering the middle class are simply too high.

That being the case, Blanca and Guido and others like them must often rely on the indeterminate factor of luck. One night in 2015, Blanca texted me with the latest development: Guido had been accepted into the William Cullen Bryant High School in Astoria. It is a decent school, about three miles from their apartment, and it has sports teams and AP classes. By 2017, many other things had happened to Guido and Blanca. Guido's grandmother had passed away. He was still in school, but working in his off-hours at a restaurant in TriBeCa. Blanca was making a better living by working longer hours as a live-in caregiver for a family. She felt like she had lived up to what she thought she might achieve in coming to America, and I had to take her at her word. Other caregivers I had spoken to had ultimately happy outcomes, after years of separation from their family, like the baby nurse Esther—I received word that she was finally reunited for good with her children in America later in 2017.

Many others might not have been as relatively fortunate as Blanca and Guido and now the baby nurse Esther, long separated from her children. But for now, they were relatively happy. For Blanca and Guido, it seemed, the indeterminate factor had come through.

6

UBER DADS

MOONLIGHTING IN THE GIG ECONOMY

Matt Barry taught history at Live Oak High School, a public school in a suburb of San Jose, California. At thirty-two, he was in his ninth year on the job. Five days a week, he stood in front of thirty-five eleventh- and twelfth-graders, instructing in AP American history and economics. But Barry also had a secret life that is becoming increasingly common among American schoolteachers—he had been spending his after-school hours and weekends as an Uber driver in order to earn enough extra money to prepare financially for the birth of his first child.

Barry and his wife, Nicole, taught, each earning $69,000. This should have placed them solidly within the middle class. If Silicon Valley had never risen to prominence, their salaries would have been just fine. But the explosion in wealth from the tech boom hoisted housing costs in the Bay Area well beyond the reach of longtime working- and middle-class residents. In Barry's neighborhood of Gilroy, a 1,500-square-foot "starter home" went for $680,000. In the town where he taught, starter houses ran up to $1.5 million, ensuring that the people who spent their days educating Live Oak students would never live near them. When Barry's child was born—his wife was fourteen weeks pregnant when I spoke to them in September 2016—the family started paying an additional $6,000 in annual health-care insurance.

Barry's Uber passengers liked to imagine that teachers could easily afford to live in their community: He would shock them when he told them what his day job was as he shuttled them around ritzy Morgan Hill, where his high school was located. Between rides, he graded papers. Among teachers, he wasn't even the worst off—he and Nicole produced two incomes, and they owned their home. Even so, they were on the financial edge. "Teachers are killing themselves," he said. "I shouldn't be having to drive Uber at eight o'clock at night on a weekday. I just shut down from the mental toll: grading papers in between rides, thinking of what I could be doing instead of driving—like creating a curriculum."

It was no accident that Barry was driving for Uber. For the last two years, the company had sponsored initiatives to encourage teachers to moonlight as chauffeurs. The campaigns differed from year to year and from city to city. In 2014, its discomfiting motto was "Teachers: Driving Our Future." In 2015 in Chicago, there was a seasonal component: Uber offered teachers a summer job. To sweeten the deal in that city, Uber offered a $250 bonus that summer to any teacher who signed up to drive by a certain date and completed ten car trips. In 2016, Uber Oregon unrolled an app that tells riders when their driver is a teacher. Uber Oregon also trumpets that 3 percent of each fare goes back to a classroom, and it offers a $5,000 bonus to the schools with the most drivers, logging the most miles. Uber has promoted its teacher-driver initiative as an act of apple-pie altruism, a perfect private-sector remedy for the failures of the public sphere. The company has teacher-driver blog posts that back up this particular spin. One Uber-teacher scribe named "Lindsey" gushed: "Every day teachers are asked to do more with less, constantly faced with new challenges and limited resources. Uber opens the door for more possibilities and delivers a meaningful impact to the communities we serve."

Beneath this feel-good frame is a disconsolate reality. Fathers like Matt Barry are "asked to do more with less," not because resources are mysteriously tight, but because the public, and the politicians who represent us, don't value teachers and workers like them enough to pay them more.

This has been true since the dawn of this country's modern education system, but the consequences have grown particularly acute recently in boom zones like Silicon Valley, where the mismatch between teacher salaries and local housing costs has become ever more pronounced. In these places, wealthy residents happily shell out for custom-built houses with swimming pools and "super basements," but are rarely willing to pay higher taxes so their teachers can afford to pay rent.

The result for teachers like Barry is that they wind up having to provide, in essence, personal services for their students' families just to make ends meet. Uber has hailed this arrangement as an "opportunity" for teachers—a chance to boost their earnings while "dedicating their lives to shaping students' futures" in their teaching jobs. Teachers moonlighting as Uber drivers are supposedly a prime example of the "sharing economy" at work. Yet, stripped of its "generous" veneer, Uber's teacher-driver campaigns are also sharing in a more twisted Silicon Valley fantasy: low taxes, good schools, and teachers who drive you home after your expense-account meal with a venture capitalist!

These conglomerates are gargantuan outfits that offer

short-term, cheap services delivered by "independent" contractors. They have become hugely successful by trading labor across platforms over which workers have little to no say.

There was also a gendered element of this dark Silicon Valley fantasia. Of the dozen Uber and Lyft driver-teachers I spoke to in 2016, most were also parents, and almost all were men. (Of course, this is often true of the workers employed by these services.)

It made me wonder whether men were sometimes more willing to literally drive the extra mile to retain their class status. After all, these men were also affected by the American societal amnesia about the cost of raising a family. Both parents routinely now work more time or additional jobs or stranger hours, or all of the above. Also, the devaluing of caring professions doesn't just hurt women like those you've met in the preceding chapters. It also hurts their men.

For men probably more than for women, external failures like losing a job or losing social class standing may cause them to lose a sense of themselves as people and citizens: according to the "precarious manhood" theory, as social psychologists call it, manhood must be worked for and sustained. As the male studies scholar Michael Kimmel writes, men are raised to expect social class privilege, having internalized the sense of themselves as their family's breadwinners, but "they are waiting for a stability that never comes. These men believed that if they played their cards right, they would get the same things that their parents did, that lower-middle-class version of the American Dream," Kimmel explained to me. "And they lost the bet."

It also made sense to me that these driver-teachers, no matter their gender, were eager to hold on to their middle-class identities—in terms of, say, housing, or income—at all costs. After all, for some the loss of such an identity spells the loss of hope and faith in the future. They are stymied, not only in terms of their own prospects but also their children's. As Barbara Ehrenreich argues in *Fear of Falling*, the main anxiety of the middle class is that we don't trust our ability to re-create our own social position for our children. The upper, or "owning," class has a simple answer to this problem: inheritance and land. For the lower class, there has traditionally been another answer that they have often been forced to swallow: that they will subsist on little for their lifetime and so will their children.

The Uber teacher-driver-fathers are one element of a broader contemporary malaise I've seen again and again. If you can't reproduce the standards you were raised with for your children, you may feel like you are living in a land of broken connections.

TEACHING HAS ALWAYS BEEN A POORLY PAID PROFESSION, PARTICU-larly considering its educational requirements and its serious responsibilities. One historical reason for lower teacher pay is that early on, teaching was considered women's work and thus merely a small second income for families, according to Richard Ingersoll, a professor of education and sociology at the University of Pennsylvania.

Teachers are no longer primarily female and working to bring in their family's second income of course, but their weak pay endures as a reflection of this earlier sexism. To help make up for the shortfall, teachers have long taken on additional jobs during the summer vacation, even though most are paid on a twelve-month schedule. What's new is the depth of their desperation. Teachers who are chronically underpaid in places like Oklahoma are forced to rely on soup kitchens and food stamps, in addition to second jobs. In

Mandan, North Dakota, I got in touch with Rebecca Maloney, an elementary school teacher and single mother with three kids who had to turn to the crowdfunding site GoFundMe just to pay for a $1,000 career development class she wanted to take at the local university. Another teacher in North Dakota I spoke to was heading off to clean houses after the final school bell in order to pay her rent. Thanks to the oil boom in that state since the early 2000s, the cost of living has risen unsustainably, while first-year teachers still make about $30,000 to $32,000 a year. Meanwhile, teachers working in increasingly expensive locales like San Francisco and Chicago are forced into the gig economy or have to work other side jobs, such as bartending, to survive.

In 2016, I spoke to a number of schoolteachers besides Barry who were racking up miles as Uber chauffeurs. John Daniels, a history teacher at James Lick High School in East San Jose, California, started giving lifts for Uber in his Toyota 4Runner on Thursday and Friday nights. Anthony Arinwine, a first-grade teacher at Malcolm X Academy in San Francisco, started with the gig economy company last summer. He spent twenty hours a week giving rides in his Nissan Altima, sometimes driving until late into the night. Salaries for teachers in the San Francisco Unified School District average $65,240 a year. In fact, they have the distinction of working for the 528th lowest-paying school district out of 821 districts in California. At the same time, San Francisco boasts some of the highest rents in the state: rent for a one-bedroom apartment averages $3,500 a month compared to a statewide average monthly rental cost of $1,750, which already is not that cheap.

"My rent was increasing—the cost of utilities was going up," Arinwine explained. "My normal salary didn't have a lot left over compared to previous years." As he watched his rent for a one-bedroom apartment in the East Bay inch up and then jump higher and higher, from $1,300 to $1,500 and then finally to $2,000, he began driving for the ride company, often late into the night, just so he could continue to live in his home. Still, he told me, he eventually had to give up his place, despite driving on the side. Now he was renting a room from a good friend. "I'm not able to buy a house. I may have to move out to a state that's less expensive, unless I get married and have a dual income," said forty-six-year-old Arinwine. His parents, a civilian worker in the military and a clerk at the police department, only attended "some junior college," he said. Yet he described them as more comfortable financially dwelling in Orange County than he was living in the Bay Area.

For Uber, economically challenged teachers like Anthony Arinwine represent a dual opportunity: a marketing coup as well as a ready labor force. In recent years, Uber has been trying to appeal to its customers as the ride service that employs middle-class people who find themselves at an impasse. When then-CEO Travis Kalanick started running the company in 2010, Uber took a different sales tack, claiming that driving for the company could be a full-time job, with drivers making as much as $100,000 a year. But Uber drivers' pay stubs show that this was far from the truth, as they earned substantially less than that.

John Koopman, a former journalist, is now an Uber driver. A bald spark plug of a man, Koopman worked as a journalist for twenty-five years, the last thirteen at the *San Francisco Chronicle,* and reported on the invasion of Iraq. He was a nominee for the Pulitzer, yet he is also representative of the many middle-class professional dads who have been sucked into the gig economy and who are also irritated, even enraged, by this fact. When I first spoke with him, he was driving constantly, pulling in $43,000 annually in his hometown of San Francisco. (Today he does not drive as regularly.) When he began driving, he was a mass of road rage over his lost career and status. Then he discovered a potentially depressing new "Zen"—the nothingness that is being an Uber driver, as he put it. It was a balance borne of nihilistic resignation.

"I have a master's degree in journalism. I've been to Baghdad, I've interviewed majors and corporals," the fifty-nine-year-old father of a twenty-one-year-old son told me. "And after I was laid off, I had trouble caring about anything: I was too stubborn or too stupid," to see a counselor or pursue other forms of self-help. "I worked at a strip club. And now I drive an Uber car. People ask me, 'Don't you want to get back in the game?' I answer, 'Not really. Who cares? You don't care.' No one cares if I am still a journalist. It's only, I have to be paid enough money to pay for my bills."

At first, he was worried that his young son and, yes, his mother would judge him for his career transformation into a strip club manager. I met Koopman at a New York café—he had driven up the East Coast in the car he used for Ubering, after having relocated to Delaware, and while I spoke to him, I thought about how my generation was lulled into believing that our lives would follow a rising arc. This belief was also held by Koopman, whose mother was a nurse and whose father was a truck driver. He was the first person in his family to attend graduate school. But as we talked, he recounted a litany of formerly middle-class parents like himself who now drove for Uber or rented their homes on Airbnb while living in single rooms of their Bay Area apartments or houses, due to the cost of living in the area.

Koopman worked extra and unlikely jobs to keep paying the bills, which included towing drunken strip club patrons out of the dance area and carrying in his arms a stripper who had overdosed to an ambulance. Koopman highlighted for me that today no amount of education can guarantee that a person will sustain their chosen professional identity, or even an identity that fits. He also personified some degree of shame. And shame is feeling one has few options: Emmanuel Levinas writes, shame is being stuck inside our own being, literally and philosophically trapped in our own skin.

"Uber has been extremely clever at finding occupations like teaching—they say that this is your neighbor driving for extra money," said Steven Hill, author of *Raw Deal: How the "Uber Economy" and Runaway Capitalism Are Screwing American Workers*. "It is all part of an Uber public relations initiative," he told me, "and the company's new narrative."

I spoke to Michael Amodeo, an Uber spokesperson, while he was walking his dog, wanting to understand more about the company's intentions with its middle-class moonlighting jobs programs. As he was busy, he cheerfully directed me to an essay by David Plouffe, on the site Medium, that echoes Hill's assessment of Uber's PR strategy but gives it a civic-minded tint. Plouffe was Barack Obama's campaign manager and an Uber senior vice president until 2015, when the company decided to emphasize that it was coming to the aid of America's squeezed middle class. The Uber platform, Plouffe writes, serves as the "pay raise they [the drivers] have not received in their other jobs." Of course, working additional hours is not a "pay raise"—like so many typical statements of the Master Gigsters (my own mash-up of "gig" and "gangster"), Plouffe's rhetoric was infuriatingly, and giddily, Orwellian.

It's not hard for Uber to acquire the appearance of authenticity and community it craves by recruiting middle-class people as drivers. In cities across the United States, there is a large schism between the cost of living and wages, especially in our most expensive places, like Washington, D.C., New York City, and Los Angeles. Uber's Amodeo, responding to my questions about teacher-drivers, later emailed me, "What we've learned is that teachers and educators see Uber as a flexible way to make money with their car." But these teachers, like nurses and other members of the Middle Precariat, are being stressed to the breaking point, and the practice of working a second job—like the teachers in Texas who told me they deliver pizzas on the side because their pay is so low—is now widespread.

Although ostensibly middle-class historically, teachers have few of the advantages that we used to associate with that standing. (Public radio programs, for instance, may choose to present a family with two teacher parents as a wholesome and unostentatious version of the middle-class life.) Teachers may not be able to afford their mortgage payments or their kids' day care, and they usually can't afford the status markers of the past—summer vacations, a car for each adult, health and retirement savings, college educations for their kids, or gym memberships. They also can't afford professional help (which they might have been able to use to discuss their unpaid bills). Indeed, some teachers occasionally must resort to Supplemental Nutrition Assistance Program (SNAP) benefits (food stamps) and other federal benefits.

HOW CAN WE ASSIST THE TEACHER FATHERS WHO MUST DRIVE FOR Uber on the side? Many of the solutions I arrived at can feel like stopgaps. They're useful for short-term coping, but they don't bring about the necessary changes on the federal and state levels. For instance, a generous system would subsidize care for their children. But there must be more granular help.

In terms of paying the rent or the mortgage, the stumbling block for some middle-class people is that the eligibility for affordable housing is defined by the median household income in a given region: that is, to qualify for rental subsidies, a family needs to earn at or below 50 percent of the median income in their area. (That number is very different in Iowa City and San Francisco.)

In recent years, a handful of local governments, mostly in cities, have begun to act on the high cost of dwelling. For instance, some places have created housing assistance programs to attract schoolteachers like the ones you have just read about, as well as firefighters, police officers, and other public servants. Chicago, for instance, recently introduced an initiative to assist public safety officers in buying homes. In Los Angeles, Milwaukee, and Hertford, North Carolina, among other places, school districts have invested in faculty housing to help teachers afford housing where they would otherwise be priced out. In 2013, Newark, New Jersey, opened Teachers Village—a six-building complex that includes residential housing marketed to school instructors.

But what are we to do, though, when a robust safety net seems out of reach? We can reach for some of the modifications that intellectuals believe can make gig economy companies like Uber more equitable for teacher-drivers and for everyone else. For instance, one utopian answer is a heady new digital phenomenon called "platform cooperativism." This mouthful of a term, now in vogue among academics, refers to taking the tools of traditional for-profit online platforms and directing them to more collaborative and democratic ends.

While they are small-bore now, these new app-reliant online co-ops also employ middle-class workers and may put more to work that they have control over in the future. For instance, Stocksy is a successful stock-photo collective that ensures that photographers are paid for their work, and Loconomics, a San Francisco co-op, hopes to compete with the "freelance labor" company TaskRabbit. (When TaskRabbit was purchased by IKEA, I had to wonder whether the ultimate aim was to provide anxious and cheap human workers to help customers assemble their Swedish plywood desks, in a batteries-included scenario.) Stocksy and Loconomics help us imagine new arrangements that could begin to address the many binds in which gig economy workers currently find themselves. The dream is for tech-savvy co-ops to empower workers; otherwise, workers could dehumanizingly be ordered up like pizzas.

As those suffering from unstable hours, precarious jobs, and low wages today include everyone from contract worker Uber drivers to document review lawyers to retail employees, platform cooperativism is now being mentioned by politicians—among them, the British Labour Party's Jeremy Corbyn, who praised the "cooperative ownership of digital platforms"—as a way forward in an exploitative age.

"If we are all producing value," said Sean Ansanelli, a platform cooperativist app developer, explaining the philosophy behind it, "why shouldn't we all share the benefits of that value?"

There are also platforms for employee-run and employee-owned cooperatives for nannies and cleaners, like Beyond Care Childcare Cooperative, a child-care co-op in Brooklyn, New York, which is run by the nanny members themselves, with the guidance of the nonprofit Cooperative Development Program at the Center for Family Life (CFL). When I reported on it, Beyond Care's thirty-eight workers were also "owners," with thirty-one clients. Members advertised their services together and pay co-op dues; in return, they own their agency and thus have more job protection and greater solidarity. Such groups, I discovered, are part of a larger worker cooperative movement in America: over 150 new worker-owned co-ops have come into being since 2000, according to the Democracy at Work Institute. Beyond Care and co-ops like it hope to provide a socially responsible challenge to gig economy cleaning companies like Handy, a domestic work platform, or Care.com, an online marketplace that helps millions of parents locate caregivers but is not worker-owned or worker-directed.

The New School scholar Trebor Scholz, a platform cooperative expert, told me that he sees the tech-enabled co-op movement as "the antivirus to Ayn Rand." In the aftermath of the 2008 downturn, with every third worker a freelancer, contingency and contract work became the norm for the middle class as well. Advocates of platform cooperativism hope that their movement can make the harsh, unpredictable gig labor market a little more humane. The hope is that consumers—maybe the types who sip fair-trade coffee—will support co-ops, disrupting the gig economy business-as-usual. The cleaner apps of the past, in the words of one organizer, were "a faceless yellow dismembered hand that cleans your house for you." The new apps and platforms can train consumers to do more than simply look for the cheapest service and the highest ratings. But what they cannot do is provide this class of worker with things like benefits, which is a big part of their struggle.

The workers now TaskRabbit-ing or Uber-ing, however, aren't thinking about high-minded conception platform cooperativism when they gigged to pay their bills. Matt Barry, for instance, mostly thought about the financial pressure of the area where he lived. Like other Uber drivers, he was plagued with guilt over having to take a side job and not spending time developing his wisdom and skill as a teacher. He thought about this when he taught the economics of Silicon Valley to his students, knowing that many of them wouldn't be able to afford to live in the area themselves when they grew up.

"I definitely thought being a teacher would elevate me to solidly middle-class and settled," said Anthony Arinwine, whose teaching salary of $70,000 a year should have given him a comfortable middle-class cushion. "I could not afford to have a child on my income: I can't afford a home to raise them," said Arinwine sadly. "I mean, I can't imagine giving kids what they need to be happy."

Arinwine told me, "I thought I'd already be not worrying about money by now. I thought I'd be making my way to retirement."

Little of this desperation comes across in Uber's teacher campaign. Instead, Uber's website teems with profiles of middle-class "UberEducators" who work as rideshare drivers to collect extra spending money. The site features Monique, a schoolteacher in New Orleans with "12 years of teaching experience under her belt" who turned to Uber to "help during the 2015 holiday season." Another driver profiled on the company's site is a frustrated special education teacher who was sick of doing "paperwork" rather than working with children. She also wanted to pay for a new porch for her home.

These testimonies, sometimes accompanied by inspirational slo-mo videos of real-life driver-teachers, fit nicely with Uber's portrait of their educators as wholesome, hardworking professionals. But they also telegraph another useful message: Uber's drivers have turned to rideshare-driving not as a full-time professional pursuit but as a second job that serves a purely supplementary purpose. This is a useful message for Uber, which has been working hard to push the idea that its 400,000 drivers are independent contractors, not employees of the company. As employees, they would be entitled to minimum wage, overtime pay, benefits, and basic employee protections—which would strike at the very engine of the Uber business model, costing the company billions of dollars. As independent contractors, they receive none of these benefits.

Uber is fueled in part by those trying to make ends meet in an overall economy that devalues their work. The sharing economy helps deny its participants basic worker rights or, in the case of Airbnb, creates more scarcity in the rental market by turning apartments into illegal hotels, in a cruel cycle.

Many drivers, along with leading labor advocates, disagree with the company's classification of its drivers as independent contractors and have begun challenging the company in court. Uber has been fending off nearly a dozen lawsuits alleging that it has misclassified its drivers, including a sprawling class-action suit filed on behalf of drivers in California and Massachusetts. (Uber settled with the California and Massachusetts drivers in 2016, promising $100 million in exchange for the right to continue classifying the drivers as contractors, but later that year the judge rejected the settlement.) When I asked Uber's Amodeo whether Uber drivers were workers or contractors, his emailed answer was curt: "People who drive with Uber are independent contractors." And then, like a winsome copywriter trying to sell precariousness itself, he wrote: "They value their independence—the freedom to push a button rather than punch a clock."

It remains to be seen whether Uber, currently valued at $69 billion, will get its way. What is clear is that, amid serious accusations of labor violations, the teacher-drivers in Uber's fleet provide a valuable veneer of altruism and authenticity. Scrolling through Uber's PR materials in praise of its moonlighting middle-class drivers reminded me of *A Modest Proposal*'s famous "suggestion" that the poor Irish survive by selling their children as food to the rich. But Uber's publicists are neither Jonathan Swift nor Juvenal. In the symbolic realm where a tech company like Uber dwells, the teacher, like the nurse or the firefighter, is instead traded, with the tap of an app, for "well-meaning" capital.

The gig economy's workers often seem to exist in the abstract, as if TaskRabbit's workers really were the cartoon bunny on its logo, or Uber drivers were simply a human-shaped extension of the company's letter U.

Today the larger problem of undervalued—and underpaid—teachers is that their years of advanced degrees and hard work are more cherished by companies trying to project legitimacy than by the politicians who have long paid teachers mere lip service. "It should be a warning sign to us that teachers have to grade papers between giving lifts," said Richard Ingersoll, his voice rising. "Just look at high-test-scoring Asian nations, where teachers are in the top of their college classes and are well paid. How do we compare?"

Invisible culprits, from the lost power of unions to the rising price of real estate to the expansion of temporary and gig labor, are responsible in large part for these individual hardships. Yet the members of the Middle Precariat can't help but blame themselves. Indeed, I started to see self-blame as a symptom of financial uncertainty, along with depression and, in some cases, aggression.

When I caught up with Matt Barry, the history teacher and expectant father, at the end of the summer, he said that he was not driving anymore. "You don't make much money in the summer, and the surges have gone down," he explained. Since he and his wife were expecting their first child, however, he had come up with another way to make a little extra—he and Nicole had just rented their house to the golf caddies for the Women's Open. In other words, he had moved from one extra gig, Uber, on top of what should have been a full-time job, to another, renting his property on Airbnb—all in a determined effort to simply stay afloat in the threatening shadows of Silicon Valley wealth.

7

THE SECOND ACT INDUSTRY

OR THE MIDLIFE DO-OVER MYTH

In a Boston classroom in autumn, rows of students were dressed up in the corporate equivalent of their Sunday best, a patchwork of different epochs of business wear: flats and beige hose, a mustard-yellow dress suit with embroidery, white dress shirts, and reading glasses. These students were not in their teens or twenties, though, but in middle age. They weren't learning a traditional curriculum either. Instead, a passel of women who called themselves "career navigators" were instructing them in the art of job interviews. If a job interview didn't land them a job, exhorted one bespectacled navigator, flashing a warm toothy smile and clad in a pencil skirt, "don't blame yourselves!"

Each attendee had paid $20 to learn things that might sound self-evident on first blush: crafting a LinkedIn profile, mastering interview techniques, and fighting the blues at mini-workshops with titles like "Battling Negativity." In the other rooms, a professional photographer was taking corporate headshots of the participants. They sat somewhat awkwardly, one by one, under the hyper-lit white umbrellas. These were middle-aged and middle-class people, a mix of white, Asian American, and black. Most were battle-hardened by job loss and underemployment. They were asking the career navigators how to get—and keep—jobs. They believed that maybe they still had a chance for a do-over, a new start. They *had* to have another career or they'd fall into penury. That many of them were parents made their quests even more urgent.

We were all at the RE:Launch conference, held by the Boston-based nonprofit JVS, or Jewish Vocational Services.

What hurdles were they trying to overcome, the career navigators asked.

"The world has evolved beyond me," answered Tamara Spencer, a former aeronautics engineer in her early fifties who was dressed, like the others, in a job-interview-ready cotton jacket. "I don't know any technical engineering for this millennium."

"I haven't worked in seventeen years," said another woman, in a reedy voice. She looked at her hands. "I've taken care of my family, but I *am* a trained lawyer. I am anticipating rejection from employers." A pre-K teacher, gray-haired and soft-spoken, confessed that she kept losing teaching jobs within a year. A computer programmer said that he was a negative person whose inner voice was murmuring, *This is never going to work out*. (I had assumed that coding is an entirely safe profession, but I soon realized that it's one that pushes out older people, fixated as it is on youth and the latest in technological know-how.) A former restaurant general manager and sommelier was also out of work: he confessed that he had just lost the lease to his apartment and was homeless.

The career navigator said that she could help all of them.

"We promise not to just tell you to just be happy or to smile," she said.

The restaurant general manager had managed to maintain an exquisite appearance; wearing a blazer and striped European designer socks, he reminded me of the central character in a French film I had seen years ago, who pretended to go to work each day in finance, lying to his family after he'd been laid off. I thought of this film as the restaurant GM told the class about his problems. Unlike many of the other assembled job-seekers, who were middle-class and unemployed, the GM would routinely get hired. Then he'd just as predictably get fired, usually after only three months on the job. "As soon as I finish their wine lists, they let me go," he said. And the owners would get their money's worth: all along, they'd been planning not to keep him, he said.

The others, seated behind white plastic classroom-style tables, shook their heads. It was September 2016, eight years after the financial collapse that had written this particular script. The bad times had supposedly passed, and employment was rising. But none of the plentiful new jobs had reached the attendees at this all-day conference. The JVS office was a renovated rabbit warren in Boston's business district: each day it was filled with the clatter of hundreds of young job-seekers, most of them working on their résumés.

"Move away from this negativity: your inner nature is not that defective person you think you are," a career navigator said. She sought to put us on a regular schedule—not working nine to five can be quite depressing, she remarked.

If a lot of other inspirational programs seem to push at least a small (and sometimes not so small) element of self-blame for the job-seekers, RE:Launch was offering something quite novel. The career navigators were gingerly sidestepping what I kept thinking of as "unemployment-shaming"; instead, they emphasized the job-seeker's self-care while hunting for work.

"Yeah, it's my mom's fault," one of the job-seekers said in a broad Boston accent, joking about his predicament.

It turned out that student or educational training debt had put many of the attendees in their predicaments. This made sense to me. After all, more than 60 percent of Americans with student debt are over thirty. While some flippant aristos like to claim, as one real estate developer did, that if millennials stopped eating avocado toast they'd be able to buy houses, a paper by the Federal Reserve Bank of New York in 2017 showed that this myth of profligate spending on things like gourmet food is simply untrue. Rather, declining homeownership rates are partly the result of rising public tuition and student debt, which afflicts younger Americans starting to have families.

At least 17.5 percent of the $1.3 trillion of outstanding student loan debt belongs to those over fifty years old as of 2015. Since 2005, there has been a substantial increase in the percentage of debt held by the late-middle-aged. Some of the increase may be due to more people going back to school for further education, but a lot of it has to do with how much longer it now takes to pay off original college debt and the rising costs of college and graduate school. The way the middle-aged folks I spoke to talked about student debt reminded me of film noir—student debt plagued them as a real-life example of a narrative being rigged, of evil shadows cast on the wall. There was a haunting, almost Grand Guignol quality to how debt collection companies chased former students like the aspiring middle-aged nurse and mother I spoke to. Some collection agents even threatened these middle-aged debtors: the world may have seemed to them to be closing in on them. The enemy here isn't human, however, but an abstraction: the numerical figure for financial expenses that have often accrued over decades.

No clever spin or special cream can hide how much age matters when trying to develop a "second act." It gets worse the older you are: if you are over fifty-five and unemployed, it will be harder for you to find a job than it was when you were thirty, according to a study on long-term unemployment by the Federal Reserve Bank of St. Louis. Furthermore, the Equal Employment Opportunity Commission (EEOC) has seen markedly more age discrimination claims in the last twenty years: annually, about five thousand more claims were made in the last few years than in the late 1990s. According to a 2014 Bureau of Labor Statistics report, long-term unemployment rates for people over fifty-five years old are around double the rates for young people. Ageism is clearly out there. If some of us are to survive as workers, we have to deny, on some level, the existence of our bodies—bodies that age and give birth.

In another era, some of the workers at the RE:Launch conference, unfolding their reading glasses and opening their laptops as if it were the first day of a school they never thought they would have to attend, would have been retiring rather than looking for jobs. Or, if their work had not been automated or downsized, they could have kept doing the work they had been pursuing all along.

"The big picture in our economy is that people are in increasingly precarious situations so they *have* to have a second act or a third act—this includes the upper-middle class," said Ofer Sharone, a University of Massachusetts Amherst sociologist and the cofounder of the Institute for Career Transitions. "At the same time, it's an extremely difficult job market for older workers and people out of work."

We in America believe in starting over—and over, and

over. But now, as the middle class contracts, these tales of costly second acts must be given a hard look. The for-profit universities, certificate programs, and coaches getting in on the new trend—all of them "helping," for a fee—add up to a larger and sometimes troubling phenomenon. I call it the "second act industry."

THE ILLUSORY FAITH IN SECOND ACTS IS NOT JUST THE CREATION OF the greedy for-profit law schools that lure anxious and sometimes underqualified middle-aged students into enrolling and taking on debt. It is *also* peddled by a whole constellation of cheery websites, special programs, self-help books, and specialized gurus. The requirement that workers sell themselves as commodities—hello, American obsession with self-promotion!—is baked in.

Coaches for the "career renegades" preach the human potential of those who in previous generations might have been planning their retirements. I scrolled for hours through inspirational accounts, talks that exhort people my age to get up off the couch and start doing what we love, that old canard. Ann Rankowitz, founder of Second Act Coaching, promises, "Your own Second Act can be your most freeing, empowering moment of your life." At the very high end, some companies guide professionals age fifty-plus on their second acts to the tune of $20,000 to $90,000 a year, counseling them on how to present themselves to new employers and what careers to "transition" to. Articles entice users to hit the "restart" button after fifty or "reboot" after forty. Websites assure us that "by learning the essential ways to focus on your added value, you can turn your age into an advantage." One coach offers "A Dose of Work+Life Confidence" and even hilariously dubs "Confidence" the true

"Vitamin C." One coach calls himself a "career designer." There are virtual summits, like Mega Reinvention and Age Busters Power Summit.

Of course, these many exhortations to brand, network, and reeducate yourself at a later age—to have your best second or third incarnation—don't arise in an abyss. There's a real demand for these services in a country whose inhabitants have been taught to seek only individual solutions for problems that are often systemic in nature. (We will learn about some collective fixes to these dilemmas later in this book.)

The professional second act arises also out of a broader faith in human perfectibility: we have a makeover mentality about identity. Finally, there is our now postmodern understanding of a life trajectory. At worst, the makeover mentality that suffuses the second act industry is like plastic surgery for careers. But changing a job or a profession doesn't work as reliably as dermal fillers, especially given our unstable working conditions and the unstoppable rise of automation now affecting white-collar careers. The idea that if we tinker with ourselves long enough we will eventually discover how to fit ourselves into the economy is a happy fantasy that doesn't always come true. The premise is flawed: you can be perfectly trained and highly competent, but if no one is hiring, you are still out of luck.

In the middle of the twentieth century, people often worked at one company for the majority, even the entirety, of their career. By the time I entered the recessionary job market of the 1990s (a time when young people assumed the identity "slacker" partly in preemptive self-protection from the rejection of increasingly fractured careers), a lifelong job at a single company was already becoming quaint. It was an antique cliché, like the gold watch given to retiring workers, a marker of a humdrum existence that had come to seem unobtainable.

People were expected to jump from company to company or to collate freelance hustles. I learned this lesson early—I even remember a professor at graduate school explaining this to me as I kept asking, plaintively, why security like my parents had was eluding me, why I had these strange little jobs writing for databases instead. The truth was that the gig and freelance economy had started to consume other jobs, and soon fewer employees would have the option to actually work for a company at all.

The second act industry traffics in such fantasies. It's a positive-thinking wonderland promising later-life second chances. The hidden driver of the industry is the loss by some people of their narrative arc. When the American Dream functioned properly, an individual might have access to a clearer personal story line. Your first job led to a second job, usually in the same profession, and if everything worked out right, you kept that job through to your retirement, which arrived at age sixty-five. This is no longer how it works, and the life trajectories of American workers, including many of the people I met for this book, are no longer as structured as the lives of their parents.

The key is for the squeezed people of America to understand that feeling this way *isn't just their personal problem.* They are being pinched by a huge system error.

GENERALLY, THE FOR-PROFIT COLLEGES AND COACHES TARGETING unemployed older workers—who are going for the second act not knowing what else to do but blame themselves and try to solve their own problems—are marketed for maximum revenue. Sharone said: "They inflate the promise of what you might get from all of these training workshops, books, coaching, or schools." They promote the notion that it's never too late, and you're never too old. "There's the tendency for these coaches to tell their students that if they follow the formula," says Sharone, "if they do everything the workshop suggests, the only thing holding them back is themselves."

Tamara Spencer, the fiftysomething aeronautics engineer, started out as an engineer in Southern California. She then left the workforce to raise her two daughters, now both in their early twenties. When her daughters were in middle school, Spencer returned to the work world.

Due to her difficulty finding a job—she said that she simply couldn't break back into engineering, as her earlier technological education had become outmoded—she became an unwilling dilettante. A petite red-haired woman with a tart and self-effacing good humor, she laid out her collage of a trajectory. First she was a part-time engineer. Then she was a website designer. "Personal chef" had seemed like a good idea; she believed that it was an easier subfield to break into. So she went to culinary school for $40,000. Her culinary school, Silver Top, "didn't coerce me" to enroll, she said, but "it was a for-profit college. It was like going into a used-car lot, and they are pushing you and pushing you and giving you these promises."

Soon Spencer was in debt for her education, like so many others, and hadn't been hired as a chef . . . or anything else. Because cooking professionally was her third or fourth profession, not the one she had been originally trained for, Spencer didn't have the years of fine dining experience that clients or restaurants tend to require and didn't get the positions she applied for.

Spencer also developed issues with her back that required surgery; eventually, she physically couldn't work in a restaurant. "It was my fault for thinking this school would be a good idea. It was the culinary school's fault for trying to make the sale. I've tried to reinvent myself so many times. To be honest, it hasn't worked."

She told me this as she made her favorite recipe, a chocolate oil-and-vinegar cake that she served in mugs.

"I feel extremely guilty for putting the burden of debt on our family," she told me. "There are psychological effects of competing with the college grads. Not getting discouraged is very hard when I am in that negative spiral, looking at all I need to get done and what a fraud I am."

Her parents were more secure than she is, even though they were teachers; her father worked as a bandleader on weekends and as a barber in the summertime. They had a comfortable upper-middle-class life and took vacations. One-quarter of their salary went to housing costs and other expenses, and they didn't have cable bills.

Of course, Spencer's parents had typically had greater job longevity and, with no need for second acts, had not been preyed upon by the second act industry. That industry's for-profit vocational schools and commercial universities are propped up by government grants. While career-training colleges have been accused of exploiting veterans and nontraditional (read: usually older) students through aggressive recruiting, they depend, like traditional nonprofit colleges, on their students' federal student aid.

Middle-aged job applicants of color I spoke to also suspected that race—and racism—was very much part of the equation when they couldn't obtain their societally promised second acts. Courtenay Edelhart, a former newspaper reporter who is black, told me how she had applied for hundreds of public relations jobs after she left her job under threat of being downsized, to no avail. She thought that her age, fifty, was part of the problem, and perhaps her race. "Being black and chubby and middle-aged? No one wants to hire you," she said. She's not all wrong. According to a 2017 study, by Northwestern University, Harvard, and the Institute for Social Research in Norway, white applicants receive 36 percent more callbacks for jobs than equally qualified African Americans.

And overall, hiring discrimination hasn't declined much since 1990.

Not getting the jobs one applies for can create a cascade of new events with lasting consequences. In Edelhart's case, in 2015, she decided to go back to college to become a paralegal at the not-for-profit College of the Canyons, where she was pursuing her associate's degree. She had one class left to go as of our last contact (and would subsequently find work as a paralegal at a law firm). In the interim, with every emergency she encountered going on her credit card—a dead car battery, a flat tire, bedbug extermination—she was now deep in debt as well.

Edelhart yearned to take her kids on a vacation—they'd never been on one. Though her fifteen-year-old daughter was athletically gifted—she ran track and played soccer and basketball—Edelhart couldn't afford to take her to tournaments or pay coaches or even put up the $500 registration fee for club soccer. Her daughter eventually was able to play soccer when the registration fee was waived for her, even though her main sport was track. Now that her daughter was a sophomore, Edelhart was trying to get her on a team by her junior year, when colleges would start recruiting, in the hope that her daughter's athletic record would help her obtain a scholarship.

"It's not her talent that's holding her back," Edelhart said, in a self-blaming mode that I knew was likely misplaced. "It's me."

It made me want to break my journalist role and argue with her and disagree vehemently. *No,* I wanted to say, *it's not just you. It's the end of journalism as we know it and it's also the pretty obvious limits of these midlife career makeovers.*

The second act industry does not aim its pitch only at individuals who are desperate, dreaming unlikely dreams, or naive, but a host of institutions do target older job-seekers by exploiting any desperation they're feeling or dreams they may be trying to fulfill. For-profit universities, like the now-defunct chain school ITT, sometimes promise educational and professional results for their graduates that never happen. ITT graduate Mynor Rodriguez, a thirty-nine-year-old father of four who lives on the northwest side of Chicago, told me his story of for-profit woe, and it was indeed really, really bad. Before he went to ITT, Rodriguez had been working as a high school–educated graphic designer. After he was "enticed" into enrolling by a visit to a nearby suburban ITT campus, where admissions staff told him that they'd help him find a job upon graduation, he signed up; eventually he obtained a bachelor's degree in information security systems, but went into serious debt in the process. His fellow students were mostly in their thirties and forties, he said. The ITT representatives pitched the program to older students predominantly.

Now, eight years later, he was $59,000 in debt to a school that was under investigation by the Obama administration. Rodriguez's level of debt and the fact that he was struggling with it are far from his problem as well. Student debt has been found to be particularly difficult for middle-class borrowers of color. As a 2016 report from the Center for American Progress and the Washington Center for Equitable Growth explained, student loan delinquency is higher in African American– or Latino-dominated neighborhoods. A further look at the data showed that the population of people of color who had the hardest time paying back their student debt was not low-income: they were middle-class students like Rodriguez. Partly because, as was the case for Rodriguez, students of color with education debt included a good number who attended for-profit schools. As graduates of for-profits, they were less likely to pay back their debts.

The for-profit-college-to-lifelong-debt scenario was certainly happening at that bad second act repository, ITT. The school shut down suddenly on September 6, 2016, stranding its forty thousand current students and leaving its thousands of graduates wondering about the value of their degrees.

Rodriguez had another problem caused by ITT. In the years after graduating, Rodriguez applied for jobs doing system administration and computer networking, including his dream job working at the Chicago Police Department, but he was disqualified from working there because CPD didn't consider ITT a valid college.

"I did this for my family, to make a bigger income and to show them an example that no matter what age, you can go to school," he told me. This turned out to not be the case. Rodriguez now makes $50,000 doing technical work for the municipality, merely $5,000 more (not adjusted for inflation) than he did before attending ITT. He now has debt to contend with. "I still can't achieve the American Dream," he said.

The president America elected in 2016 had a surprisingly large hand in one of these for-profit scams. At Trump University, many middle-aged Americans paid up to $35,000 to attend a "school" that in reality was not a credentialed university and that eventually faced numerous lawsuits for misleading students. Perhaps it's fitting that the institution's namesake was a man who endlessly reinvented himself at older and older ages, moving on from each failure as if it had never happened, until he finally, shockingly, became president and had a golden financial cushion. The claims—over conning students into believing not only that they were going to a real university but also that they would have access to Donald Trump himself—ended with a $25 million settlement. The outcome for Trump University clearly shows, yet again, how rotten some of these institutions are, as well as how high up this particular form of corruption can be found. The myth of educational second chances can be worse than misleading—swallowing it whole can be disastrous. This industry is another example of how the system is rigged against an unstable middle class. After all, even a U.S. president has a huge stake in this fraudulent system.

For-profit universities aren't alone in taking advantage of those seeking a second act career. Also getting in on the scam was the investor Armando Montelongo of San Antonio, Texas, who ran sham real estate "flipping" seminars. One hundred and sixty-four middle-aged former "students" joined in a 2016 lawsuit against him, accusing the sometime–television presenter of sapping their life savings and even driving one to suicide.

But it's not just the obvious scam artists who are a problem. Some of these experts are legitimate, just not always able to help.

The legitimate counselors include people like "career pivot" expert Marc Miller, who explained to me how he aided people with their second acts. His clients were typically fiftysomethings who had been recently laid off. When

I first spoke to him, by phone, Miller told me that he charged $1,500 for an evaluation and then $3,000 to $5,000 to help package his underemployed or unhappily employed charges. (He later told me that these prices were "ballpark.")

The second act that Miller was peddling differed from the career conjured by many for-profit colleges, graduate programs, and online MBA programs. Miller told his clients that a master's degree was a waste of time. "All they've done," he exclaimed, referring to these graduate students, "is accumulate student loan debt!" Miller favored instead the professional certificate.

Miller first gives his clients a personality test, similar to a Birkman behavioral assessment, an occupational and social questionnaire. Then he decides whether they need to be "branded or rebranded." Miller next writes his clients' "brand story"—because, he told me, "the worst person to write your 'brand story' is you."

He attributed some of his skills to his temperament, describing himself as a "relational kind of guy" who could "network better than anyone."

He also offered his own biography as a testimonial. At sixty-one, he lived in a paid-off condo, had a "small seven-figure portfolio," and hoped in the near future to sell his second act coaching business for a substantial sum. That will be his fifth act, he said, after an earlier career in the tech business.

THE VISION FOR SALE IN SECOND ACT INDUSTRY FEEDS DIRECTLY on deep currents of U.S. social history.

In the late nineteenth century, the concept of failure changed. It became "modern, a low hum rather than a loud crash. It meant a fragmented life, not necessarily a shattered one," writes the historian Scott Sandage in his history *Born Losers.*

A few decades later, there was a new faith in startling ascents and turnabouts, in our ability as Americans to bounce back from failures. As F. Scott Fitzgerald said of this alchemy, "I once thought that there were no second acts in American lives, but there was certainly to be a second act to New York's boom days." (The incorrect, commonly misremembered version is "There are no second acts in American lives.") Like other glam modernists, and in the midst of the Depression no less, Fitzgerald affirmed second acts and their metaphoric power. The novels about failure's manic alter ego, class mobility, are full of this sort of admittedly doomy reinvention: Fitzgerald's *Great Gatsby,* Saul Bellow's *Augie March* ("I am an American, Chicago born—Chicago, that somber city—and go at things as I have taught myself, free-style, and will make the record in my own way: first to knock, first admitted; sometimes an innocent knock, sometimes a not so innocent"), Theodore Dreiser's *Sister Carrie* and *The Financier* ("A real man—a financier—is never a tool. He used tools. He created. He led."). The fiction of the last century was about monumental zigzags—people going from nothings to "somebodies" to different and sometimes tragic somebodies.

By the 1950s, literature defined failure by stuckness—people frozen in place by their origins, incomes, and jobs—and many protagonists were incapable of personal renovation. Arthur Miller's Willy Loman in *Death of a Salesman,* for instance, embodied the American Everyman Failure.

In the 1960s, self-help cultic organizations and movements, like the Human Potential Movement (HPM), further rebuked the self-image of the Lomans of this country. HPM's apostles believed that human possibility lies fallow in most people because they don't develop and unleash their creativity. The "ruthless self-centeredness" of HPM, in the words of Barbara Ehrenreich and Deirdre English, represented how the marketplace had braided itself into the most personal relationship we have: our relationship with ourselves.

As the sixties flowed into the seventies, that faith in human potential became more and more corporate. Today it is echoed by self-help books for striving professionals who happen also to be foundering. The virtual bookshelves heave with titles like *Life Reimagined: Discovering Your New Life Possibilities* (on the high end) and *The Real Brass Ring: Change Your Life Course Now* and *Boundless Potential: Transform Your Brain, Unleash Your Talents, Reinvent Your Work in Midlife and Beyond* (on the cornpone low end). These tomes encourage frustrated middle-aged readers, in the words of an interview with one of the self-help authors, to identify "specific steps and formulas for effortless life reinvention." For instance, the author of *The Real Brass Ring*, Dianne Bischoff James, was thrilled to embark on her own second act at thirty-eight after a life-changing encounter with a psychic. James writes of herself, quoting the psychic's words: "Dianne, you are a talented writer, healer, teacher and performer. But sadly your life is heading down the wrong path. Your brass ring is coming by and you need to grab it before it's too late." (Perhaps the most extreme example of the second act mentality in action is Donald Trump's presidency, which features a protagonist who compromises himself and everyone around him in order to stay in power and stay rich.)

And that is the warning so vibrantly conveyed in the sales pitch of the second act industry: *Act now! Time is running out!* There is some shaming energy there too: if you don't act, you have no one but yourself to blame. Warning and shaming give way to the golden promise: if you *do* act, imagine the new horizons, the riches.

"I've found that a lot of the consultants for middle-aged job-seekers are all about teaching marketing," said JVS career counselor Amy Mazur, a white-haired denizen of the Boston suburbs whose crinkly smile, sensible clothing, and ethnic necklace reminded me of a few northeastern therapists I had previously encountered.

In this sense, said Mazur, what job-seekers need is an intervention not only in their own lives but also in the coaching and counseling industry that caters to them. She saw their problems not as a failure to brand, but rather as more existential and political and based in solitude, ageism, or bias. In the workshop, she coaxed job-seekers to talk about their emotions, including sadness, with her and each other.

There's a lot of emotion involved in trying to reinvent yourself at midlife, much of it arising from self-blame. Most of the second act-ers I spoke to tended to blame themselves rather than the true authors of their ills: the costly graduate schools that got them into deep debt; the former employers who kicked them to the curb; the bias against mothers and older workers. I saw the attitude of self-blame among the Uber-driving teachers as well. They first put the onus for their dilemma on themselves. They assumed that they needed to create their own self-directed transformations. These drivers also didn't expect organizations, such as unions or churches, to help them as they might have done in the past. As the Columbia Law School labor specialist Mark Barenberg writes, that is partly because America is experiencing the lowest levels of union enrollment since the 1920s—the era before industrial unionism began to rise in the 1930s. He attributes this attrition, which he calls a "crisis," to managers retaliating against workers collectively organizing as well as a political shift to the right. Deregulation and globalization are also culprits. In my own field, I saw publications like Gothamist and DNAinfo shut down in 2017 after employees tried to unionize (the organization I run even created a special fund to help these now-unemployed journalists).

When I spoke to squeezed parents I saw the personal effects of this. I saw the degradation of workers' rights in professors like Brianne Bolin, nurses, and retail salespeople—the more unstable hours; the greater use of short-term contracts; the reliance on more complex and longer day-care hours than were optimal for their families; and of course, the less coherent professional arcs, as jobs without union protection became harder to keep into middle age.

Like pregnant workers who try to downplay or hide their pregnant state, second act-ers are also being asked to deny their biology.

ONE BIG BIG WAY TO COUNTER THE SECOND ACT MYTHS WOULD BE to curb federal funding for for-profit schools, like the failed ITT chain or the colleges associated with Corinthian. Their degrees can be less likely to lead to full employment, yet may cost the same as or more than degrees from public universities or community colleges.

Taking another tack, activist group the Debt Collective organized what they called the "Rolling Jubilee" which erased $17 million of more than 12,000 students' loans by buying their debts from collection agencies for pennies on the dollar, with funds they raised from multiple donors. Belle Goldman, for instance, came home to her Los Angeles studio apartment, patted her cat, and tore open a letter in a plain white envelope. The way things had been going, she told me, she'd expected to find something unwanted inside:

a scam offer perhaps, or yet another unpaid bill. Instead, she found a note informing her that she was no longer in debt for around $1,500 for the two months she attended Everest College. "It sounded too good to be true," Goldman said.

It seemed to her like a Ponzi scheme in reverse. But both the horrors of student debt and the few feel-good rescues like this one are real.

The Debt Collective had another solution for those who have wound up in extreme debt to for-profit colleges and universities. Under the collective's aegis, former students organized in 2015 to go on debt strike and refuse to pay back their loans from the for-profit college Corinthian. (In 2015, the Consumer Financial Protection Bureau won a $500 million lawsuit against Corinthian, an award that seemed to support the striking students' perspective on Corinthian.) In 2016, the Debt Collective organized a similar debt strike against ITT. By early 2018 the strikes had won millions of dollars of relief for defrauded students. (This month, my alma mater, Brown University, launched another idea on the front end: raising enough money for financial aid to prevent students from ever having to take out loans in the first place.)

Curbing or minimizing federal subsidies for the unscrupulous for-profit schools is the first correction for the second act industry, and encouraging healthy skepticism toward costly certificates and coaches is another. But the final correction is a psychological and internal shift within underemployed middle-aged workers themselves. Sometimes those I spoke with felt like they had been forgotten, perhaps even discriminated against. Jobs programs must focus on bolstering workers' low sense of self-worth.

We can also look carefully at what goes *right* when people succeed, for clues. After working as a librarian but barely keeping her head above the floodwaters of penury, Michelle

Belmont, whom we met earlier in this book, found that her economic misery had somewhat receded. The hard, sour center of her problem remained her debts, much of them for graduate school, as was true for so many other middle-class parents I spoke with: as of 2017, Belmont still carried an enormous $20,000 in credit card debt and $175,000 in student loan debt. But she had made life shifts that nonetheless seriously helped. Her family was now renting a house in a cheaper area. "Maybe once I get my credit score up, we'll be able to buy something, in five years or so," she projected. (As of our last contact in 2017, she had achieved her goal even faster than hoped and was closing on a house that year.) In addition, her son's new day-care center was a welcome surprise, "wonderfully" inexpensive and "awesome quality," she said.

One reason why Belmont was able to turn her life around was that she tried to build her emotional resilience. Bolstering one's "grit" is what the American Psychological Association (APA) recommends to combat money-related suffering (it seemed wrong to me to assume that those experiencing economic insecurity are not resilient in the first place, or that all they need is more grit, a greater ability to "bootstrap" toward security, rather than, say, better jobs or adequate and affordable day care for their children.)

In Belmont's case, she also focused on practical things like realistic goals and "small accomplishments," as the APA suggests, such as updating her résumé, rather than just on emotional abstractions. She also took "decisive actions," another APA recommendation, rather than, in the organization's words, detaching completely from her problems. One of those small but decisive actions was "ambitiously" job-hunting, as she put it, until she found a new full-time gig that paid at least $100,000; with her husband's $55,000 income, her new job put them very solidly in the middle class.

She had been making $37,000, but when she looked for jobs again, she discovered that her skills were worth more than that.

"It's such a relief off my shoulders to not feel like I have to be ashamed at having depression and anxiety, and getting stuck in a cycle of debt," she said. "I still occasionally borrow money from my family for groceries, but I'm always able to pay them back within a month."

Belmont's story is one of the relatively rare second act success stories. She recognized that her position was not her fault or her spouse's fault. The couple is now financially far more resolved.

Her unexpected conclusion was also evidence to me of the way narrative reversals—the ones that contemporary Hollywood screenwriters orchestrate—sometimes happen to people in real life, though the happy reversal is often smaller and subtler. Things get worse, but they also get better. Turnarounds happen, but they are modest, neither tragic nor triumphant.

AT THE BOSTON RE:LAUNCH CONFERENCE, THE CADRE OF UNEMployed men and women were led, thankfully, toward solidarity and acceptance of their own fragility. Some no longer had a clear, meaningful, or even reasonably profitable way of earning a living, and certainly the markers of an ample middle-class life were still out of reach for many of them. But the "career navigators" neither coaxed them to self-brand nor hectored them, like football coaches trying to get players to try harder.

That acceptance of the would-be job candidates came through at the conference when the navigators went over "tough interview questions." "The question I am dreading in an interview is: 'Why do you have a big gap in employment?'" said one jobless participant. A tall and husky bearded fellow raised his hand. "What about when they ask me, 'Why have you been underemployed?' Do I say I have a disability? Should I tell them that I have clinical depression?"

The navigators suggested that he stress another element of his "narrative" besides his years ensnared by massive melancholy. He had invented a new computer language, he said. Should he mention that instead?

"I am looking for your professional self: everyone has a professional self," said the navigator, optimistically.

The navigators went over actual interview techniques. A finance worker originally from Hong Kong, with a young son, asked the navigators how to present herself.

"Don't let silence in an interview intimidate you," said a navigator. "When you go in, talk for ninety seconds or two minutes to start with about yourself, no more."

"Admit you don't know the work culture at the company."

"Can I lie about my age?" called out one woman, giggling.

"You can't hide that!" a navigator, clad in a long-sleeved dress covered with large pink tea roses, replied. Then three navigators explained, in chorus, how easily any job applicant might get caught out falsifying their age—there are now digital records of all individuals, as we live in a surveillance society enamored with coded archives. Perhaps they should direct attention elsewhere—for example, by focusing on putting words like "traveler" and "specialist" on their LinkedIn profile near mentions of the more quotidian professions "hospitality" and "dentistry."

At the beginning of one RE:Launch class, one out-of-work man told the workshop how he vented his professional frustrations to his boyfriend, and another participant said that she resorted to binge eating. At the end of the event, though, they both said they felt better. Even the quasi-homeless erstwhile restaurant manager murmured, "I feel pumped from these discussions.

"Now I can only be let down when I don't get hired," he added, with a minor note of mirth. The roomful of other job-seekers of a certain age laughed, albeit grimly, the sound of bitter warmth.

8

SQUEEZED HOUSES

On a sunny autumn afternoon in a bohemian commuter village in Dutchess County, New York, two children, aged four and five, were playing in their family's living room. The girls had just finished a lunch of cold pasta, cheese sticks, and homemade, heart-shaped juice pops. They were now taking turns on the yoga swing, a shiny contraption made of cloth that hung from the ceiling. Inevitably, they began to argue.

"She's not giving me my turn!" one girl shouted.

"Will you get off the swing?" the other girl's mother, Mary, murmured gently to her daughter.

"No, no, no!" her daughter, Nona, answered, angrily.

"She said she doesn't like me!" Astra protested. (The names of these families have been changed to reflect privacy concerns.)

Eventually, the first girl's mother, Jennifer, started a countdown and asked the girls to relax. She then told Nona to get off the swing. Triumphantly, Astra got on the swing and rocked back and forth with great gusto. The other girl began to cry.

The two mothers then attempted to talk the situation

through with their daughters. "Why are you crying?" Mary, who had a dark brown bob and the pacific gaze of a professional yogi, asked the girls. The mothers then suggested that the girls make peace with each other by playing instead on the outdoor trampoline. They agreed to share it, as they shared almost everything.

The women were coparenting. The term describes parents who have never been linked romantically or biologically yet live together and parent their children collectively. (It also extends to more conventional situations, like divorced or separated parents caring for their children in a time-split arrangement.) Parents like Jennifer and Mary coparent in part to save on the cost of rent or a mortgage and the high cost of child care. For some people, the very concept, with its inherent questioning of personal space and traditional roles, might cause unease for a wide variety of reasons. But for others, coparenting might be the ultimate fantasy—a way to have a kind of extended family, without the enforced obligations of romantic or blood relations, but with the bonds created by chosen solidarity.

Mary and her daughter lived in the same home with Jennifer, who was thirty-six; Jennifer's husband; and the couple's daughter. When I first encountered them, the families had met only two months earlier, after which they decided to enter into a coparenting arrangement. For the three parents, the relationship had no romantic component; rather, it was based on a mixture of idealism and pragmatism. While coparenting arrangements may sound a little like the communes of yore, they're a lot less hippie-minded and more about financial security than earlier communities. Real estate values in Jennifer and Mary's town had been skyrocketing for the past five years. Jennifer, who sometimes worked as a graphic artist, and her husband had bought their one thousand-square-foot house for $270,000. It was on the edge of what the family could afford, said Jennifer, so they had agreed to offset costs by renting the basement level. The couple also believed in sharing child care, an idea to which Jennifer was drawn partly in reaction to her own childhood. Her parents had frequently left her with sitters while they worked sixty-hour weeks, and she wanted to do things differently. "I couldn't live without another family now—I wouldn't know how to," she said. "I like the transparency of living in a community. There is no place to hide: people know my business."

The coparenting arrangement with Mary in a riverside town was the third such one that Jennifer and her husband have had. They previously lived in a nearby apartment building that was fairly communelike, with three other families with children, plus a few single and coupled adults. When the interpersonal dynamics eventually grew too difficult, the couple bought their own house. They shared it for a time with their best friends and their daughter, who later moved to the Boston area for work. So Jennifer and her husband placed a Craigslist ad seeking someone to share in a coparenting arrangement.

The house included a two-bedroom basement apartment, which they listed for $1,200 a month, with an offer to reduce the cost for the "right person." (Apartments of comparable size elsewhere in town went for $1,300 to $3,000 a month.) "There are energetic types I work with and ones I don't," Jennifer said, adding that she'd interviewed many people before finding Mary. "I like people who are introverted, sensitive, and communicate well, who respect their own boundaries as well as other people's. We wanted to actively coparent with a parent with a child or children our children's age." They ultimately agreed that Mary would pay $1,000 a month in rent, plus electric. In addition to dividing the house and child-care duties, the families shared many meals, bills, and a car. Although the setup is inherently lopsided, given the owner-tenant relationship, the living arrangements and the shared responsibilities are meant to be equitable.

Mary told me that for her coparenting was less a matter of philosophy than of economics. Life for her in this arrangement was now less expensive. She'd left Astra's father, whom she described as belligerent, when she was five months pregnant. After her daughter was born, mother and child had slept on the foldout couch in the living room of her mother's place. Mary worked retail in a specialty food store, relying at times on food stamps for her own meals.

Four years later, Mary and her daughter had an altogether different life. Their apartment had the piney simplicity of a yoga center, with clean wood floors, a guitar hanging from the wall near an antique instrument, and an abstract magnetic sculpture that her daughter made. Fall foliage and a neighbor's apple tree were visible through a window. Mary was working at a fancy day spa; she credited coparenting for allowing her and her daughter to have their own living space, one that she could actually afford, for the first time.

IN THE YEARS AFTER I MET MARY AND HER DAUGHTER AND ENCOUN-tered their complex arrangement, I kept running into more coparenting setups, conducted in similar or even grander registers. In later 2017, nearly two hundred miles away from Mary and Co. in Boston and its suburbs, I met two other families that had also practiced this unusual model. One had gently transitioned out of it. Sophia Boyer, an educational consultant and former teacher at both public and private schools, was its particularly silver-tongued explicator. For five years, Sophia, now in her forties, had owned a coparenting house, where the inhabitants took care of each other's children. She had raised her twins, as of this writing aged thirteen, partially, with eight or so other people, three families in all. As all the parents had different work schedules, if someone worked in the morning, another parent in the group might drive a kid to a doctor's or dental appointment. The parents were first- or second-generation immigrants, including Sophia, who was born in Haiti. Others derived from Guatemala and Senegal.

The roughly 3,800-square-foot home in Brockton they shared was painted yellow, and it was big enough for "people to go off on their own," said Sophia. It included a third floor with an "in-law" suite. The dwellers rarely had to worry about cooking for their own children on their own either. Though they didn't have a formal schedule, they'd organize meals and activities. First, they'd shop and cook food for large groups—pasta, rice, meats from Costco, seasonal platters, Haitian food, Guatemalan food. The ingredients had to be inexpensive and healthy, and the person least employed that week would take the lead on the meals. The kids would play before eating. Then they'd dine, as if they were a single biological clan.

Overall, the families lived far more cheaply than they would if they had been carrying all of their own rents or mortgages or day-care bills, or even filling their own pantries all on their own. It was, Sophia said, a "replacement of upper-middle-class support and trappings" that these families were lacking.

Sophia felt that her background, growing up partly in Haiti, had made the coparenting paradigm seem normal and was part of why she and the other parents, from immigrant families, shared the same "philosophy of child-rearing." This "philosophy" (outside of the coparenting) was actually quite traditional. "Children did children stuff, adults did adult things," Sophia said. "It's a first-generation-attitude."

Sophia and the other parents were also motivated by the stark economic fragility of being a middle-class parent in America. Boston and its surrounding areas had become too expensive for any of them to easily purchase homes, even though they were all working professionals at the time, with bachelor's and master's degrees. "We were middle-class on paper but we didn't have financial backing of our families because they didn't have enough time in the country to accumulate wealth—no one in our extended families could give us $25,000," said Sophia.

This was, of course, a pattern I saw over and over again for the squeezed middle-class parents of color I interviewed, whether immigrants or not. Their families of origin did not have as much savings as most of the white middle-class parents. They were less likely to own property, in part due to far less intergenerational family wealth. Those whose parents did own real estate may have possessed homes in black neighborhoods that didn't gain as much value as white ones. (One study attributed this lag to white home buyers rejecting black neighborhood and lowering housing demand.) Behind it all was America's history of racism.

Sophia said she made about $48,000 a year at the time and sometimes less. At certain points, her salary was the same as the private school tuition at the school where she worked. The other parents in the house—social workers, accountants, graduate students—were also earning $45,000 or $50,000 with one making $30,000. They tended to believe in "ownership but not hoarding," as Sophia put it. She and her then-husband rented rooms to the other coparents "for very little or nothing at all."

As money was tight, the whole group passed along old kids' clothes, for instance, all organized by size and gender, so families wouldn't have to buy new threads.

Sophia considered the arrangement as nearly idyllic and strangely tensionless.

Eventually, however, Sophia and her husband divorced—they sold the house. They and the other coparents scattered.

Sophia's acquaintances still carry out a similar alternative parenting survival strategy. Mea Johnson, for instance, is part of another active coparenting house in Boston. I met Mea at an event in a Jamaica Plain church where I also met Sophia.

Mea calls herself "indigenous"—Apache—and she wanted her son to be raised in a community of families of color. So she moved with her child to the coparenting Black Indian Inn for five years, an indigenous black house. Now Mea lives in the Margaret Moseley Cooperative, which is "predominantly composed of people of color," as she told me: four white people, thirteen adults, and six kids. The idea, Mea said, was for the coparenting home to avoid "replicating the system that already exists" and instead to share equity in the house itself, making up for generations of disenfranchised people in America not owning property (and of course also having their property stolen from them at some point in their ancestry). A single mother and community organizer like Mea, without inherited wealth, would otherwise be unlikely to find financial security or homeownership in today's America. Coparenting was one of the few and relatively radical tools available to her.

COPARENTING AND OTHER INGENIOUS EXPERIMENTS, HOWEVER, can seem merely clever pebbles thrown against the Goliath of an absence of federally supported day care or easily accessible affordable housing. Such measures would significantly ease the financial stress of parenting. But in their absence, with families like Mary's left in the lurch, bespoke solutions like coparenting are gaining currency.

"Lots of families are sharing outside of the traditional unit. It's increasingly common," the scholar Kathleen Gerson said. She also pointed out that in disadvantaged families, the aunts, grandmothers, and other extended family members who are often called in to care for kids in a pinch may live in-house.

Gerson, a professor of sociology at New York University who specializes in families and the new economy, told me that one of the stumbling blocks to making such cooperative solutions possible is that Americans are wrongly "obsessed with the idea of the nuclear household. Family is not just people related through law or family ties, but those who come together to share care, and provision of economic resources." So coparenting has long been a strategy that people use "under conditions of hardship," she said. Like day care or college or graduate education, housing has become so expensive in some areas that it's one of the key ways in which families get squeezed. Many of the parents in this book were either barely clinging to their homes or being pushed out of the cities they had long dwelled in owing to the expense and scarcity of reasonably priced housing. Coparenting was one small way they managed.

After all, due to the scarcity of rentable or buyable housing in many cities, it is now common for teachers and other professionals living in places like San Francisco to seek second jobs, like the Uber drivers we met previously. As Princeton sociologist and demographer Douglas Massey has written, many families living in costly American cities or surrounding suburbs and towns are feeling the effects of the rise of the digital economy on the cost of real estate. With the rise of that sector, well-paid tech workers moved into urban centers, altering the composition of these areas. The companies themselves sometimes moved in as well, forming mini-metropolises within already au courant places, as Amazon has done with its headquarters in Seattle. The working poor and the struggling middle class then increasingly move to the outskirts—the outer parts of boroughs or the inner suburbs or the uncool far-flung commuter towns like Vallejo, which is thirty-two miles (via a backbreaking commute) from San Francisco. Fashionable cities are now intensely striated, best explained with phrases like "spatial inequality" that reveal social class segregation (as well as racial segregation), rather than intermingling.

Among the many middle-class strivers who have literally been unsettled from their neighborhoods and cities in the last decade—and are living out class segregation—is Courtenay Edelhart, whom we met in the previous chapter. In her former career as a business reporter, Edelhart, now fifty, *specialized* in writing about the financial hardships of others. Before it happened to her, that is.

After twenty-five years as a newspaper reporter after obtaining a degree from Northwestern University, Edelhart was out of a job. Once she lived in Bakersfield, California, where she reported for the *Bakersfield Californian*. There she was initially making $46,000 a year. That wasn't really enough for a single mother of two to afford to live in one of the most expensive states in the country, but she stuck it out because she loved journalism. Seven years later, after annual reductions of either salary or benefits, she was down to $40,000 a year. Finally, toward the end of her stint there, her salary had been reduced to $39,000, all of her friends had been laid off from the newspaper, and she was made to understand that she was next. The 2008 financial crisis and recession exacerbated unprecedented income inequality and left a bitterly divided nation, one vulnerable to "alternative facts" and, ultimately, the election of Donald Trump. It was not an accident that this was accompanied by immense reductions of newsroom staffs, especially in rural areas far outside the media elite quadrants of Washington, D.C., and New York City; up to 40 percent of those employed in journalism were laid off. Edelhart was part of that contraction.

By then, Bakersfield had become unaffordable. When Edelhart was priced out, she moved to a more remote town, but the move increased her commuting costs, in both time and money. The family relocated to an apartment in the desert city of Lancaster, an hour north of L.A., where a three-bedroom place rented for as little as $1,085; it also happened to be closer to school. Edelhart paid roughly $1,200 monthly for her place. In Bakersfield, her family had been part of a community they could rely on for last-minute child care or rides when the car broke down; now the family of three was more cut off.

Even with lower real estate costs, Edelhart's family was forced into a more ordinary version of the coparenting situation: they acquired a part-time roommate, another former reporter in her forties, who was also pursuing an associate's degree in paralegal studies, at a nearby community college. Having a lodger—even one who was a generous classmate who paid to stay with them three nights a week while commuting to school—was a matter of necessity for Edelhart's family. Edelhart was, after all, an overqualified underemployed student. Even after finally landing a job at a law firm, she was still stretched thin financially and burdened by extreme debt, both from credit cards and student loans. "I am probably going to be in debt for the rest of my life," she said. A single mother, she couldn't afford her twelve-year-old son's ADHD medication, so he had to use older versions of what doctors had prescribed for him in the past, medications that were covered under Medi-Cal. These were less effective and had more side effects, said Edelhart, but he had to take them because she couldn't pay out of pocket for the newer drugs her plan didn't cover.

The end of so many newspapers has led to epic layoffs and just as many attempts at self-reinvention. What is most striking, however, and frequently left out of the "end of journalism" stories, is that reporters like Edelhart are emblematic of the crushed middle class of the future, not just precious anomalies mourned by a self-absorbed media class. Her mother and father, who had triumphed over their own working-class backgrounds to obtain master's degrees and become social workers, were much more solidly well off than their daughter would ever be, but they didn't have much left over for her to live on, either in their wills or to support her while she went back to school. Her father had since passed away, and her mother had moved in with Edelhart's family.

The downside to the family's move to Lancaster was not just social alienation. It was also the limits on the Edelhart children's freedom in their new neighborhood. "My kids have cell phones, but I am afraid for them to be outside alone—we live in a crappy neighborhood," she said, adding hopefully, albeit plaintively, "right now."

Edelhart couldn't help but take her situation personally. She knew that being middle-aged and African American might be part of why employers with implicit (or even explicit) biases had not hired her. But it was hard for her to not personalize her condition, even though for me it was clearly also social and political. After all, white applicants are a third more likely to get asked for first-round interviews than black applicants. Still, she said, "I feel like such a failure," speaking fast.

When Edelhart was a journalist in the last half of the twentieth century, it was easier to cover rent or mortgages in cities because property values had dropped. This trend reversed itself, though, when those same depressed values encouraged investor risk-taking in urban real estate and cities grew thick with new office buildings and restored housing. By 2000, middle-class families were priced out of Boston and Philadelphia and other cities famed for their knowledge production, such as San Francisco and New York City. People with power congregated there, demanding restaurants and galleries, while longtime residents watched openmouthed as housing costs went way up. Today, as the urban scholar David "DJ" Madden wrote in a book coauthored with Peter Marcuse, "real estate is attacking housing," with the "pursuit of profit in housing coming into conflict with its use for living." Like the Uber drivers, middle-class people who now want to live in or near desirable cities may need to work second jobs. This is what the working poor have long done, of course, to stay afloat.

One solution is to broaden rent stabilization, a system that permits a middle class to stay and flourish in expensive cities. I grew up in such a rent-controlled apartment—a book-lined prewar with a sunken living room and a roach problem—that cost far below market rate. Until this year, I lived in a similarly book-lined rent-stabilized unit that has glowing Ashcan School views of water towers to go along with the apartment's silverfish.

Rent stabilization and control go along with better-regulated real estate development overall, especially in desirable cities.

Such fixes, of course, are easier said than done.

There are also smaller, artisanal solutions like cohousing: people living in multigenerational communities of private, single-family homes gathered around communal spaces, or "common houses." Like coparenting, cohousing arose partly to address the alienation of the nuclear family, the financial slippage facing middle-class families, and the growing unaffordability of housing in attractive urban areas. For more than two decades, my uncle and his family have lived in a cohousing community in Washington State, one among the 160 such communities that now exist in this country, with another 130 communities in formation. An invigorating sideline development is called "retrofit cohousing": changing a neighborhood collectively over time as opposed to purposely building new housing to be cohousing.

Other off-label survival strategies include one of the oldest in the world: barter and trade. A forty-two-year-old labor organizer I spoke to in Rochester, New York, for instance, had come to depend monetarily and personally on a parenting subculture that exchanges day care for other services. It's a DIY system in which money does not change hands. When her daughter Zaya was four years old, the woman, Carly Fox, earned a little too much—$42,000—to be eligible for a government subsidy or a voucher to help offset day-care costs, so she cobbled together day care utilizing her mother, her father, a paid cousin, a neighborhood sitter, and time swaps with other parents (sitting for each other's kids). Fox had a master's degree from Cornell University, yet she had also been on Medicaid and food stamps. She was partially able to afford relatively cheap day care and housing because of her city's depressed economy. Rochester, home to Kodak (no longer as profitable after the digitalization of photography), is now a white-collar ghost town lined with gracious and underpriced homes.

In a sense, the techniques used by Fox and others to hack the cost of living as a parent by depending on their neighbors is part of a time-honored, traditional way to pool parenting responsibilities among families. The stress on single-couple families, each confined to its own dwelling, arose only after the industrial revolution: before that, families often lived within extended families on farms where everyone worked together, in a fashion that has been romanticized by poets and historians. We might argue that middle-class and wealthy children have enjoyed their new familial privacy, as it was accompanied by inventions like high chairs and nurseries. Working-class children only lost out, however, as family and caretaking networks splintered. Relying on immediate family, such as grandparents, for child-care assistance may be a bearable solution. But few people in my own peer group, for instance, have extended family who live close enough, or are willing enough to help, to make a difference. Caregiving has become more and more walled up and private as the lack of shared services has been turned into a desirable fetish—upscale parents emphasizing the pristine and individually tailored "one-on-one" care provided by their children's nannies and tutors. Yet many online cooperative services and for-profit businesses have emerged in the last decade or so to meet the demand that private-care networks, such as CoAbode, a website that allows single mothers to locate similarly situated parents for house-sharing, cannot.

Such arrangements have complications, as you might expect. For instance, I spoke with two of three women living as coparents of two children of two different mothers in Oakland. One told me that she would certainly not be able to afford a private home or regular babysitting if all three weren't going in on it together. "We are all active participants in discipline and don't have a long-term commitment to each other."

The strains in such a relationship became apparent, however. With each interview or interview request, the considerable friction between the three coparents appeared to sharpen, mostly having to do with real or perceived racial, cultural, and social class differences among the women. I received scathing emails about white privilege and communal living from one of the members of this group—who said that, because she earned less, she was asked to do more chores and day care than her coparents. I even wondered whether my questions hastened their breakup: within a year, one of them had a new coparenting situation. As I spoke to these former coparents, I felt that their underlying problem, shared with so many, was the larger struggle to survive as American families. Why haven't state-based or federal policy fixes been implemented that would free us from our exhausting and often unsustainable independence?

For me, part of what is appealing about collective parenting solutions goes beyond money and the cost of day care and housing. These arrangements can also lessen what I have found to be the abject loneliness of parenting today. I remember feeling emotionally wracked, physically stitched, and undone when my daughter was very small. How do I follow an infant's cues? When is she crying in pain? When is she crying because that's her job? Tears are the work of infants, I learned. "The Princess and the Pea" is actually a story of infancy: the princess is a baby, of course. I learned my baby, though, and I also *learned from* her, including how different she was from me and the other adults I knew. Her babyhood was probably the hardest job I'd ever have. What if I'd had someone to share this job with besides my husband, and the very loose, far-flung, and sparse extended family we retained? What if someone else had lived with us, or within hearing range? How much better and easier would it have all been? What would it have been like to raise her in one of the communities where coparenting is common? I was not drawn to this unusual solution from economic desperation or even the same level of instability as some of the coparents. I can't say I understand Mary or Sophia Boyer's situations firsthand. But I yearned to experience the warmth that radiated out of their descriptions in first person. I was also impressed by how they had taken the pattern of their lives into their own hands.

After all, as parents, so little feels entirely in our control, and financial shakiness only makes it worse—taking *anything* into our own hands can make one start to feel better about one's lot. The patron saint of second wave feminists, the poet Adrienne Rich called motherhood a state of both "power and powerlessness." Sure, you can feel powerful—or at least something like a blotchy inner glow—when you manage somehow to pursue professional achievement and nurture a squalling child all in the same day. But you can also feel extremely lonely while serving your child dinner on your own. Then there is the scraping of uneaten mac and cheese off her plate, the forcing a comb through her knotted hair while she kicks and screams, having your focus broken into pieces that never fully cohere again. In addition, the expectations of motherhood encircled me and many others, guilting us about late pickups or for buying rather than making our children's costumes or relying on frozen chicken nuggets rather than cooking from scratch.

And of course, anxiety about money thrummed under this loneliness and guilt.

"WE ARE SO ISOLATED AND LONELY AND FEEL LIKE WE DON'T HAVE community," Sophia Boyer had explained to me, a few weeks after I first met her. She was speaking broadly of parenting in America but I felt like she was talking directly to my private self. "We buy stuff to fill us but that stuff doesn't do it."

Sophia had moved out of her coparenting arrangement, alighting with her girls in an apartment in Cambridge's swankier Harvard Square. But still she attributes her daughters' continuing "good attitude," as she puts it, to their early collective living. "They learned the art of sharing young."

With the phrase "the art of sharing," I remembered my visit to Mary and Jennifer's home two years previous. The afternoon light was fading and the air was cooling to autumnal. The adults moved out into the garden to talk. The girls kept playing inside, with the father loosely supervising from his home office. The mothers recapped the yoga-swing kerfuffle: their takeaway was that it wasn't serious. They also talked about the house and their plans to grow more food at home. Mary had gardened in a plot within a huge community garden near her mother's house, and she'd had to leave it behind with the move. Now she brightened at the thought of planting a garden at her new home.

Coparenting can be challenging for children, who are asked to share their toys and their living space, transportation, kitchens, and bathrooms. Throughout the day, one of the daughters kept muttering that she needed "alone time."

Both mothers had grown up in families that felt separated from other, with little interaction among neighbors.

Mary's mother had raised three daughters on her own, without a husband or community support—and without much broader systemic aid, for that matter.

They saw their version of family as a correction of all that. And indeed as I packed up my bag to leave, the friction that had played on between the two girls all day subsided. Astra was singing the alphabet song and sticking silver and gold stars to her face. Her young friend—and housemate—was doing the same.

9

THE RISE OF 1 PERCENT TELEVISION

I spent a week with the financial adviser Marty Byrde, who, while under threat of death from a Mexican drug cartel, had been forced to launder millions of dollars in a rural red state. I then passed a few hours with an uneducated, entitled young queen, her racist, callow husband, and, finally, a twisted hedge fund billionaire.

These thumbnail sketches may conjure a gallery of villains. Yet these are descriptions of heroes of some of the most highly regarded television and streaming series of the last five years.

I was watching them on television, of course, along with millions of other Americans. I like to watch what I think of as "aspirational" television, or "1 percent TV": televised narratives about a 1 percent (or close to it) that acts with impunity.

One percent TV divides into two camps. The first camp centers on brilliant, albeit extremely violent, entrepreneurs. Our antiheroes have technical specialties that they have managed to turn into criminal know-how: on *Ozark*, a Netflix show, money management becomes money laundering, and on *Breaking Bad*, high school chemistry instruction becomes meth production. But these heroes also experience some of the challenges faced by viewers. They have had their livelihoods threatened. They are betrayed by their coworkers. These shows subtly argue that their protagonists have been forced to become criminals to avoid falling out of the upper-middle class. They are, after all, self-made: they have no rich grandparents or parents to bail them out. These series are far from real, of course, but they do rest on the bedrock American reality of income inequality—the huge gap between the hedge fund masters of the universe, such as Axe on the show *Billions* or even *Ozark*'s Marty Byrde, and the "little people."

Then there is the other kind of 1 percent TV, where the televised wealthy are more ethereal than these driven and violent aspirants with no moral code. Unlike the hardworking industrialists and tycoons of the past, the heroes of these shows are wealthy because they are tastemakers, branding experts, investors, or inexplicably well-born. They, of course, reflect the fact that those at the very apex of the apex of the 1 percent in the United States have received the greatest gains, usually through investment rather than labor and through a tax code and political system rigged in their favor.

I decided to explore these shows in these pages partly because I myself had turned to them when I was hiding my own financial instability, when pregnant with my daughter. I still use them as a painkiller in this fashion today. These fantasy narratives have created other worlds for many viewers like myself and given us a sense of continuity in a time when many of us are unstable professionally, with no idea of what will happen tomorrow. I want to escape into them, to judge these characters for their moral laxity while most likely living vicariously through them. I started doing this in 2010, when my squeezed work and home lives ran in parallel to an expansive, simulated world on the screen. What was popular back then was the 1 percent TV heritage drama *Downton Abbey,* which featured the Crawleys, an aristocratic British family in the earlier part of the twentieth century whose dramatic entrances in frock coats and devoré dresses distracted me from my physical discomfort and scattered future. The sting of monetary deprivation drew me to these overwrought spectacles of early-twentieth-century British aristocratic life: I could almost smell their byzantine meals and the nursery where future regents were looked after by governesses while their parents partied and dined. (And I was not alone: the final episode of *Downton Abbey* drew 9.6 million American viewers.) On *Downton Abbey,* the "downstairs" servant class is unruly and often downright evil: in the first season, I watched, while pregnant, a male servant and a lady's maid conspire to harm their fellow servants and that same maid plot to have her pregnant mistress trip on a bar of soap so that she would lose the baby. In contrast, the aristocrats are benign. *Downton Abbey* turned the elaborately set tables of yesteryear's *Upstairs, Downstairs,* a PBS hit and British import that ran in the 1970s and was fonder of the inhabitants of "downstairs"—the bluff, friendly maids and cooks—than television is today.

I wasn't alone in my passion for escape into screen worlds oozing with disposable income. Others too wanted to click and swipe onto shows featuring violence, candelabras, and private helicopter rides, servants polishing their masters' silver teapots, and private rap shows for leather-clad homeboy-millionaires. The Bureau of Labor Statistics' yearly survey in 2016 found television-watching to be the most common

American leisure activity: we spent about two hours and forty-four minutes per day—or more than half of our free time—watching television.

Some of the professionally unstable people I spoke to remarked on their love for reality television about the indolent 1 percent. "It's so much easier to hide in Netflix and the like," one squeezed parent said. "Television is easy, a short-term fix to a long-term problem. I watch it for the great clothes and houses and think: *Why can't I have that?*" Online, fans who get behind 1 percent shows like *Billions* and *Empire* will fight any haters. "Those who are nitpicking this probably suffer from Rich Envy," wrote one *Billions* citizen reviewer on Amazon about any commenters who might cast aspersions on that show and its central character, a twisted billionaire. "Awesome show! You feel like you get a 'real' glimpse into an uber-intelligent (though morally flawed) super-rich hedge funder," wrote another. As one citizen reviewer of *Downton Abbey* with the fitting handle workingmom29609 put it, "It's wonderful to escape to the life of the landed rich."

I like to watch 1 percent shows because I want to watch privileged people behaving badly. Yet I desperately want to see the ultra-rich find their comeuppance too.

That comeuppance never comes.

Nevertheless, I don't turn off *Ozark* or *Billions.* I don't completely yearn for their downfall, and neither do their other fans.

This sort of lazy 1 percent TV really took off with the debut in 2004 of Donald Trump's reality show *The Apprentice,* which ran in different iterations for over a decade. On his show, Trump treated his TV employees as if they were plaintiffs in a Maoist show trial. (As Trump said at that time, "My jet's going to be in every episode.") It all got me wondering whether the more staccato and dreary our real work lives are, the more we might depend on shows featuring a colorfully and darkly immoral overclass.

One percent reality shows feature outrageously lavish weddings, maternity concierges, and planners for children's birthday parties, ridiculously roomy mansions and glorious Los Angeles modernist homes, private concerts, elaborate wardrobes, and haute couture hairstylists. In one section of 1 percent TV sit the Kardashians and the *Real Housewives,* but also the 2011 show *Pregnant in Heels,* where an insanely wealthy "maternity concierge" came out about her infertility on air. "I discovered I had a heart-shaped uterus and after surgery didn't ovulate again," said the show's star, Rosie Pope, in an interview. The staggering cost of the IVF she underwent to conceive is glided over.

One percent TV goes some way toward explaining where we are politically, and it also partly explains why parents like myself and the ones in this book feel so bad about ourselves, blaming ourselves rather than a system failure for our plight.

As James Wolcott wrote in *Vanity Fair* of today's shows about the most affluent (and the invisibility of their labors), "The vast wealth depicted in movie/TV drama today is divorced from manufacturing and an invisible army of laborers doing their part; it's divorced from anything resembling work. It's floating and caressing, immortalizing."

For nearly a decade a seemingly endless assortment of 1 percent shows have appeared on Bravo, revolving around so-called housewives and realtors, who have for years composed the aspirational slice of "reality television." These characters' bodies are gym-toned and manscaped. Catty assistants schedule their every movement. These people have "people" and multimillion-dollar real estate everywhere. A decade ago, in a *New York Times* article, the construction of the "Housewives" reality television franchise was explained as a sort of "real" spin-off of *Desperate Housewives.* In that article, the scholar Pepper Schwartz coined the term "aspirational TV" to describe what we view when we ogle salmon-colored women in little heels striving to occupy the term "socialite." The lifestyles of the rich and famous have also been amply celebrated in the real estate sections of newspapers, in glamour magazines, and even by some contemporary art photographers, whose images of the 1 percent can verge on something I have called "bling porn." Bling porn—when grossly gilded worlds are turned into art that is itself grossly gilded, or when artists make luxury objects out of the luxurious lives and objects they ostensibly seek to critique.

The results of the 2016 election—as well as Trump-related television viewership numbers—point to the extension of aspirational TV into broadcasts of electoral politics and news shows as well. And indeed, President Trump is part of 1 percent TV. His media bid for president riveted millions of viewers, including those who despised him: from March to September 2016, Trump received the equivalent of about $2 billion in free media exposure, according to an analysis by mediaQuant. Much of it was negative, but the high level of television exposure helped make him president. Much of the exposure involved his extreme wealth and outrageous behavior. From the Boeing 757 jet, the gold tower where he lived, and his gauche boasts of enormous net worth to the daughters and daughters-in-law and wife with perfect and expensive dye jobs and, most likely, extensions clattering onto stage behind him, he offered voter-viewers the equivalent of a backstage pass, an entry into a sphere that seemed beyond reach for most of them. Our president and First Lady tend to routinely celebrate their own excess in photographs, as do Treasury Secretary Steve Mnuchin and his rapacious wife, who famously posed with sheets of cash.

This was part of Trump's media success—the election conceived as a way into a billionaire's life with his model wife. As Will Wilkinson wrote in the *New York Times* after Trump's election, his appearance of "majesty" helped him seem legitimate. These signs of majesty were all televised and lapped up by his faithful.

The sociologist Rachel Sherman, author of a book on the very affluent called *Uneasy Street*, told me that all of the images on television and social media of "glitzy thoughtless entitled wealth" have a negative effect, not only on those striving or treading water to stay in the middle class but also on those who are very wealthy. "The Kardashians, Rich Kids—this is what it means to be wealthy," said Sherman, speaking of the social media phenomenon Rich Kids of Instagram, which is only matched by the venal Private School Snapchats.

The extreme tackiness of these shows lets the more decorous super-rich who benefit from American inequality see "themselves as good and not that wealthy," Sherman noted. "If you are not that materialistic or Trump-y about it, you can imagine you are a better rich person, you are working hard, you aren't living in a mansion, you are setting limits for your children. But the question is not whether individual wealthy people deserve their wealth when they aren't Kardashians, but does *anyone* deserve this much wealth?" It's not fashionable today to talk about what it means to have a society we can all live in comfortably together, Sherman said, or about the causes of inequality among us, like the decline of unions. She took my "1 percent TV" idea and ran with it. "It's a thing. And also, the people who are supposed to be middle-class on television? They are actually rich."

Increasingly, Americans are watching more media than simply what is concocted in Hollywood screenwriting rooms or by splenetic political speechwriters. They are also likely to be consuming media that has been created by people like themselves—citizens posting their personal lifestyle home movies on YouTube, Twitter, Facebook, or Instagram. Although we might hope that this kind of expression would be a healthy 99 percent counter to 1 percent TV, it's often not. Instead, we tend to produce a sort of 1 percent social media about ourselves after consuming 1 percent TV about others.

Many of the Middle Precariat parents I spoke to mentioned getting a daily sense of vertigo when logging on to social media. Though this can be a positive experience—one isolated professional mother found a whole ecology of allies online—the "Yay me!" tone of many social media exchanges can also intensify loneliness and shame about your own social position. "Who needs other people's Instagram feeds that seem designed to make you feel poor, uncool, and old?" one woman remarked to me. "Facebook is the devil," another mother told me as she tried to retain her dignity when confronted by constant "Insta" photos of sun-dappled family vacations. Those trumpeting their status victories can create shame among their cohort that is not sharing in the spoils. (As the writer Michael Lerner, who worked on a study of the middle class for the National Institute of Mental Health, wrote in the *New York Times,* "I found that working people's stress is often intensified by shame at their failure to 'make it' in what they are taught is a meritocratic American economy.")

According to the CUNY Graduate Center media scholar Lev Manovich, the sense that other people's social media feeds are startlingly and often inaccurately ritzy is not an illusion. We are indeed creating aspirational visual stories about ourselves in the way we present ourselves online. With an economist, Manovich studied millions of social media images shared in New York, Bangkok, São Paulo, and London and called the result Inequaligram. Analyzing 7,442,454 public Instagram images shared in Manhattan over a five-month period, he and his economist collaborator found that Instagram images posted by locals were largely of the wealthier parts of the borough, regardless of the social class position or home location of the person taking the photo.

On Twitter, Facebook, and Instagram, many Americans now present their own timelines of self to their mini-publics each day. These tend to feature the 1 percent aesthetic—they celebrate family vacations at the sea or in the mountains, costly bronze earrings, or homemade apple crumb bars. Viewing these images makes us feel envious and less worthy, of course. We immerse ourselves in the land of those who are far more fiscally comfortable—or wish to appear that way—simultaneously escaping into the wine-on-the-beach photos of our so-called friends while hating ourselves for not being able to achieve it. This does not happen in other countries, Manovich told me.

"In a place like Moscow, the city's geographies are covered far more equally. In New York City, why are people not representing themselves in areas where they actually live?" Manovich asked. He then probed further, noting that on platforms like Instagram, people tend to prefer a minimalist visual style, an aesthetic aided by iPhones. Minimalism used to be the earmark of privilege. "These are not rich people, and they haven't gone to Harvard," said Manovich, yet they do know how to take pictures that give off an air of "economic and social prestige," as he put it. "A refined aesthetic has been appropriated by the masses."

Indeed, anyone visiting a social media platform—including

the people I interviewed, my friends, and, yes, me—can't help but notice the people projecting a look of greater money than they themselves might possess. (Through a filter, you can't tell whether a dress is acrylic or silk.) People also promote their proxies for class status—attractive spouses and adventurous vacations. These images projecting prosperity—"wealthies" rather than selfies—are a kind of extension of what I call self-branding in my book *Branded*. These personal images tend to show people radiating imagination or gleefulness or taste, or they depict families living their "best life." Mark Leary, a professor of psychology and neuroscience at Duke University, says that our social media presence tends to prop up and inflate our social position. "By posting selfies, people can keep themselves in other people's minds," he writes. "Through the clothes one wears, one's expression, staging of the physical setting, and the style of the photo, people can convey a particular public image of themselves, presumably one that they think will garner social rewards."

How can we really understand our status and standing if we are surrounded by Facebook-happy faces on one side and 1 percent TV on the other?

YOU MAY ARGUE THAT PEOPLE HAVE ALWAYS PUT THEIR BEST FACE forward for their peers. Or you may wonder, what of the wealthy of television past? For instance, what about shows like the 1980s drama *Falcon Crest,* which telegraphed the villainy of its cast of jezebels, who were certainly opulent and vile? I was raised on their shining mansions and Texas office buildings: they were my 1980s "cathode ray nipple," as the singer Gil Scott-Heron called TV. Even further back, we could consider the old masters' paintings that celebrated a new kind of wealth. But these floor-to-ceiling nineteenth-century portraits of robber barons or the opulent soap-operatic programs of my childhood presented very different rich people from the wealthy we see on TV today, in television's golden age—otherwise known as "peak TV." Peak TV shows include *The Wire, The Sopranos,* and *Mad Men.* Premium cable, with its loosened content restrictions and quality programming, has made possible a period of what the Brandeis University scholar Thomas Doherty calls "Arc TV," or "adult-minded serials" whose story lines unfold "over the life span of the series."

Historically, television has also fixated on the upwardly mobile, of course, according to Jason Mittell, a Middlebury College professor of film and media culture and American studies. Television has always turned upon what advertisers wanted to sell audiences—the lipstick that will land you the right lover, the perfect mints or deodorant to ensure that you will also be professionally appealing. When producers have occasionally tried to get working-class voices on-screen, advertisers have balked, with the exception of "little blips here and there," Mittell told me. (The flannel-shirted ressentiment of the 1990s show *Roseanne,* set in a working-class milieu, is one data point.)

But today's streaming portraits of privilege are far more violent and suffused with social class anxiety than those of my childhood. The dastardly rich of the older shows would never be mistaken for today's affluent antiheroes; the TV families of the past were long-running dynasties for whom financial ruin never reared its head.

Though we meet him after he's been threatened by the Mexican cartel, *Ozark*'s Byrde began working for that crime syndicate in order to maintain his family's plush life after his wife stayed home with the children and then couldn't get back into the workforce. This seems like a far-fetched plot until we remember, again, that middle-class life is drastically more expensive than it was twenty years ago. After all, the average top 1 percent family now earns forty times what the average family in the bottom 90 percent earns. Household incomes nationwide in 2014 were 8 percent lower than in 1999, according to Pew Research. But you would never know that from most of the serial television shows that distract us, week after week, from our own decay.

"There is never a moment that the characters in *Ozark* relax about money," Martín Zimmerman, *Ozark*'s executive story editor, told me. "The reason for this is there is no social safety net in the U.S.: once you have achieved any economic stability in this country, you live forever with a fear of falling out of it. In the show, the Byrdes can never exorcise the demon of starting out without money."

The show's rustic underclass family, the Langmores, interact frequently with the upscale on-the-lam Byrdes. Their meeting across the class divide is very important to the show's themes, says Zimmerman. In addition, he and others scripted Wendy Byrde, Marty's wife, played by the actress Laura Linney, to come "from a very working-class background in North Carolina."

In this show, the truism that we are not as likely to do as well as our parents did in the past is front and center. The family at the show's core have plenty to start with, but are now "striving not to feel unstable: they are fleeing from fear," said Zimmerman. "They have enough money to cobble together $8 million in assets to give back to the cartel while still driving a ten-year-old Camry with cloth seats," Zimmerman added. This itself is aspirational for viewers who are not wealthy: even though he's running from a murderous cartel, Byrde's life is somehow still desirable.

Christina Wayne, the former head of programming at AMC in the era of *Breaking Bad,* thinks that show's popularity had a lot to do with the recession of 2008. The show started that year but took off in 2009, at the trough of the downturn. "That was when people saw what the finance class had done to America, that the people on Wall Street were getting richer and richer," said Wayne. "They wanted to see their heroes getting back at these people."

In doing so, these characters may indeed be going beyond good and evil: they rob and fleece "rednecks" and "thugs," and they then go after people who are even wealthier than they are—such as drug dealers and rival hedge fund managers. *Ozark, Breaking Bad,* and other shows like them are also revenge fantasies, said Wayne. These antiheroes may be wealthy, but they enact vendettas against the ultra-rich. In this way, they are still part of the destabilized and resentful middle class—like their audience.

Many of the families in this book also watch a steady stream of these shows. As one middle-aged job-seeker told me, "Everyone on these TV shows is well groomed and beautiful. Meanwhile, this isn't my life! I owe stuff for my daughter's school, and rent is not cheap."

Those who watch *Billions,* as 6.3 million people did weekly in its first season in 2016—delivering "some real capital gains," as *Entertainment Weekly* put it—are watching a somewhat evil, ginger-haired Long Island mega–hedge funder whom they are meant to envy and respect, despite his evident criminality. If they watch *Empire,* they are following a richer-than-Croesus hip-hop mogul who is also most definitely a murderer. (*Empire*'s audience share is outstanding: the 2016 midseason premiere averaged 12.2 million viewers overall.)

Both seasons of *Billions* frame Bobby Axelrod, "Axe," a

handsome, redheaded hero, a man who is faithful to his wife and takes care of his friends. He can get the mega-band Metallica to perform for him privately, yet he forgoes the pleasures of a groupie's flesh. Another point in Axe's favor: he's self-made, while Paul Giamatti's character, Chuck Rhoades, is old money. What makes this positive framing particularly fascinating is that Axe is deeply ethically challenged. He wasn't at work in his firm's World Trade Center office on 9/11 because he was with his lawyer—he was about to be fired for legally murky trading. After many of his colleagues perished, Axe took over the firm. In addition, he lied about his rescue efforts—instead of saving the dying that day, he made a mint out of the tragedy by immediately shorting airline stocks and the like. In another unsavory chapter, Axe prevents a dying employee from getting an experimental drug for his cancer: for elaborate reasons, his longer life could have been a legal disaster for Axe. More generally, Axe and his workers traffic in insider tips and elaborate schemes that flirt with illegality.

Part of the appeal of *Billions* and other shows like it is the opportunity they present to escape into the lives of people at the top who are pulling the strings and refusing to play by the rules.

There are plenty of reasons, however, for the rise of today's upscale antiheroes that have little to do with America's cruel sociology. These callous binge-TV aspirants also emerged owing to trends in television production.

TV was originally considered a "feminine" medium as Michael Newman and Elana Levine have argued, one laden with housework product ads and domestic plotlines. They write in their book *Legitimating Television* that the medium attempted to become culturally legitimate by going "masculine," emphasizing damaged antiheroes who spend season after season doing battle with other men. But the aspirations of screenwriters' rooms to become literary salons don't explain fully why *Billions* and *Empire* are so popular and why *Ozark* has been quickly renewed for a second season. These shows are partly the result of a brain drain from other mediums like film and theater that helped "prestige TV" develop.

With the birth of high-end television in the late 1990s, TV showrunners were trying to remove the stain of the déclassé from the medium, often by describing their shows as a series of short films, or as "novelistic." Mittell, the television historian, told me that this sort of relativism is part and parcel with "complex TV," a narrative style in American television series that expects viewers to be constantly confused about characters' identities, about their motivations, and even about time and space. Thus, despite the murders committed by *Empire*'s Lucious Lyon in the name of his record label and *Billions*'s insider trading, the protagonists are still heroes. On *Empire,* Lyon family members murder and backstab, in giant lofts and mansions, then drink champagne like filtered water. Other shows are full of helicopters, fixers, and hot aquiline wives shtupping in swimming pools and prancing about in the latest luxury-brand peasant blouses.

These shows are also appealing because they are immersive and visually convincing, fictional simulations of the real world that viewers can lose themselves in. In fact, such shows have become even more immersive, thanks to more refined cinematography and set decor and the sheer size of the television screens on which we watch them. Consider AMC's retro "quality television" show *Mad Men:* the environment of privilege in Don Draper's mod 1960s offices and his various luxury homes was so highly detailed, from the clothing to the decor to the cars, that the art direction alone became a big part of the show's allure—far more so than was the case with the prime-time 1 percent dramas of the 1980s, such as *Dynasty*.

HOW MIGHT WE BEGIN TO CREATE AND CONSUME CULTURE THAT reflects where we are *really* at—often laden with debt or hopping between impermanent jobs? And what might doses of such bleak realism do for our general state of mind?

In fact, there is one television genre that is critical of wealthy disparity: the television comedies and dramas about less-than-regal and glossy themes, like debt and failure. This genre, which is so much smaller than 1 percent television, is "inequality entertainment."

Inequality entertainment may sound like an unlikely solution to inequity as we know it, but I see it as part of the remedy. Why? Inequality entertainment turns our self-consciousness or the money we owe into mass entertainment. Our immersive screens otherwise traffic in mostly flattering tales of the 1 percent. Inequality entertainment challenges what lies beneath some of these opulent images. It's the tens of thousands of personal student debt stories posted on social media and YouTube that tear the veil from the happy Facebook images. On television, inequality entertainment includes shows like *Mr. Robot* and *Silicon Valley*. No longer is the story of income inequity delivered via a well-meaning, crushingly earnest indie film by John Sayles, or in a single laugh line on *Roseanne*. *Mr. Robot*'s creator, Sam Esmail, drew on his personal experience with being in the red for his education. He has said that he did not finish paying off his loans until 2015, years after he graduated. *Mr. Robot*'s post-Occupy portrayal of the wealth gap follows a fragile, bug-eyed fellow named Elliot Alderson who in one season hacks his friend Angela's student debt in the hopes of easing her monetary suffering. By day he works at Allsafe, a cybersecurity company; by night he works to take down companies like it with an anarchist group similar to the guerrilla hacker collective Anonymous. Its goal is to cancel all debt.

Mr. Robot's Alderson is given to deadpan enraged voiceovers about the horrors of the contemporary human condition. "Since when did ads infect our family albums?" Alderson said. "Since when did one become greater than ninety-nine?" This is also true for HBO's *Silicon Valley,* which examines the huge gap between ramen-eating, couch-surfing, low-end tech workers and their 1 percent tech-guru overlords—the show's main source of drama and humor. The absurd tech-titan excesses in *Silicon Valley*—the venture capital toga parties, the private Kid Rock concerts—are starkly contrasted with life on the lower rungs, where a desperate aspirant pitches his app to customers at the liquor store where he works. Time and time again in the show our heroes' huge deals fall through, sending them back to penury and couch-surfing and the antic aggrieved world of 24/7 coding in airless rooms.

One percent TV, however, dominates over inequality entertainment. If 1 percent TV and indeed our president's television antics are a guide, the ultimate luxury is not just getting to watch extreme wealth, but getting to see the immensely well-off do as they wish, regardless of the consequences, unmoored from social norms, like the gods.

10

SQUEEZED BY THE ROBOTS

At a gargantuan hospital in Pittsburgh, as large as three city blocks, those staff crisscrossing the length of the building are the people you'd expect to see: nurses, doctors, pharmacists, a cleaning staff. They are joined by a small army of very different workers, though: twenty-six robots, eight in the pharmacy and eighteen working the halls. These robots "run meds," as they say at the hospital, usually making forty or fifty deliveries a day. They haul fresh linens as well as dirty ones, along with medical waste. They roll by, looking like microwaves that have spent time on a stretching rack.

These robots are not just a quirky curios docked at this single futuristic hospital, the University of Pittsburgh Medical Center Presbyterian Shadyside. They are practically commonplace at other hospitals as well: nearly 140 hospitals now employ 500 robots. TUGs, as they're called, have also worked in hospital pathology units and blood banks. The medication delivery robots are programmed to require only a biometric access and pin code from a human to finalize the meds' deliveries. Though they talk to passersby, they are usually androids of few words. "There's nothing aggressive about them—they barely say, 'Please, get out of the way,'" the spokesperson for the TUGs manufacturer told me. "They are like doormats!"

The robots don't save the hospital money . . . not yet, at least. In fact, they are exponentially pricey. But Presby and hospitals like it are embracing these cold new staffers because they believe they will eventually cut costs. I am told that the pharmacy robots supposedly save technicians' time by traversing miles of hospital floor and going from pharmacy to pharmacy on their own. This happy-speak, however, masks the reality that TUGs are being deployed to help hospitals limit the number of humans they employ. In fact, over time they may make the Middle Precariat even more precarious.

"The robots help us save on the cost of additional bodies, or FTEs," Al L'Altrelli, the administrative director for Presby's pharmacy, told me. "What are FTEs?" I asked gingerly. "Full-time equivalents," he answered.

That's a full-time *human* employee to you. At Presby, the pharmacy robots are freeing the hospital from having to hire a blend of pharmacists and pharmacy technicians—in other words, they are enabling the hospital to *not employ* highly trained human beings who make somewhere between $80,000 and $110,000 a year, plus benefits. The pharmacy workers are not unionized, he added.

Concern about the rise of the robots has become widespread, the stuff of trend pieces and hand-wringing remarks by famed techno-positiveists like Elon Musk and Bill Gates. (Gates, for instance, thought that governments could tax companies that use robots as a way to generate alternative funds for displaced human workers and pay for training in jobs that won't be replaced.) However, I've been encountering robots less as a TED Talk abstraction than as the literal professional rivals to the middle-class people I have met for

Squeezed, whose jobs may be, will be, or have been replaced by automation. Until now, many of the jobs lost have been in the automobile industry and on the factory floor. Now automation is moving into areas like nursing and truck driving. The TUGs at Presby are just a small clique of many. Robots speed along the corridors of the University of California, San Francisco's Mission Bay hospital too, prompt and servile, delivering meals and drugs to patients at a clip. A mechanical limb finds and sorts medicines. The robot arm is very, very accurate, as Aethon, the technology company that manufactures TUGs, assured me—as in never wrong, they say. And instead of needing coffee in the morning, or even wanting to talk to others, the TUGs simply charge up in a docking area.

Nurses and pharmacists, however, are simply the beginning: in other formerly secure "middle-class" professions, including the automotive industry, journalism, and law, robots are also on the path to success. The use of Daimler driverless trucks, for instance, now in a decade-long trial period, threatens human workers. The end of trucking would very likely mean suffering for real families. The American Trucking Associations says that there are 3.5 million professional truck drivers in the United States, including people who drive other delivery vehicles. Their pay tends to exceed the national average. The potential to earn $70,000 per year, with overtime, plus medical coverage, can make truck driving a blue-collar job with white-collar pay.

TUG robots and driverless trucks are far from the only encroachments on both the livelihoods of human workers and the human experience of patients and clients. They are part of a broader trend in what is sometimes called "the future of work." The World Economic Forum (WEF) in 2016 projected a total loss of 7.1 million jobs by 2020, two-thirds of which may be concentrated in office and administrative jobs in health care, advertising, public relations, broadcasting, law, and financial services. (Women's jobs account for more than five jobs lost due to our automated friends for every job gained.) The National Science Foundation is spending nearly $1 million to research a future of robotic nurses who will lift patients and bring them medicine while keeping living nurses "in the decision loop." And as a 2013 McKinsey Global Institute report on disruptive technologies explained, highly skilled workers could be put on the chopping block with the expanding "automation of knowledge work." A 2016 survey by Evans Data Corporation of 550 software developers found that 29 percent were most worried by the possibility of being replaced by artificial intelligence. Finally, a 2015 study by Ball State University showed that while trade accounted for 13 percent of lost U.S. factory jobs, robots and new technologies were the often-hidden thieves of the nearly 88 percent of the remaining lost jobs. Some high-end economists who study employment trends believe that "nonroutine cognitive tasks" will be computerized out of existence; indeed, people like Karl Fogel, a partner at Open Tech Strategies, have been known to use the hideous phrase "surplus humans."

Bank associates are being replaced by online banking, and movie attendants by computerized robot ushers. A recent report by a market research company asserts that by 2021 robots will have eliminated 6 percent of all jobs now performed by American humans, and a fair share of them are middle-class jobs, like the work of pharmacists or a certain cast of lawyers. Software has replaced workers in fields from law to tax preparation, and the latter is almost certain to become mostly automated in the future. H&R Block, one of America's largest tax preparation providers, is using Watson, IBM's AI platform, rather than the earnest, consoling human CPAs featured in the bad advertisements that played on

TV thirty years ago. Even though automated tax preparation may one day make doing your taxes cheaper, it's also ultimately another bit of code that reprograms America's middle class out of existence.

I'm reminded of the Ray Bradbury short story "There Will Come Soft Rains," about a computer-controlled house on a morning in 2026. No living residents remain in this rich and empty home after an unspecified catastrophe, perhaps a nuclear winter, but programmed toasters and robotic house cleaners still tend to the daily chores for a nonexistent family. (The story made quite an impression on me as a preteen.)

However, this potential reprogramming of the middle class partially out of existence doesn't seem to bother everyone. "We see the robot as a tool for automation," Anthony Melanson, the marketing doyen of the TUGs maker Aethon, told me. "These are worker robots."

I ask Melanson about the robots' learning capacity. They are "no Watsons," he replied, referring to IBM's famed genius computer. The TUGs simply get the job done, he added.

"Working-class robots?" I joked.

"Every reporter has an angle," he answered tartly.

I couldn't help wondering whether it's wise to replace human workers in an often upwardly mobile, skilled profession like nursing. The more I contemplated TUGs, the more I wondered whether we should be questioning the automation aficionados' fetishistic ardor. We've been to the automation rodeo before, after all. (This time the bull operates with AI.)

Shouldn't we always first and foremost defend people and their labor?

AS A TECHNO-PESSIMIST, I AM FAR FROM CHIC. BUT MY MELANCHOLY over robots puts me firmly into one of roughly four strands of thought about automation among (ostensibly human) "thought leaders," labor organizers, and the like. The first strand argues that the robots are coming and it's a horror movie for which we've only seen the trailer. The second strand also believes that the robots are coming, but that their arrival will be a techno-positive revolution! The third strand asserts that the advent of robots is both inevitable and overstated; automation may have to be addressed, according to this strand of thought, but not with great urgency. The fourth strand is certain that robots are ferociously coming for our jobs, but that there's an excellent solution for the resulting disruption, one whose intellectual trendiness and utopian excess often make people roll their eyes: universal basic income.

I subscribe to the first line of thought, which highlights what robots may bring. Some of the concerns voiced by this strand are minor and some are immense.

The Middle Precariat people I met who would soon be affected by the rise of the robots were similar to the trucker Finn Murphy, the author of the memoir *The Long Haul*. Murphy explained to me that if long hauls become autonomous, as has been threatened in the next ten years, his driver friends will most likely foreclose on their trucks. With their limited education, and being in late middle age, they'll only be able to work for places like Walmart—at best. This change is happening already. The first commercial delivery made by a self-driving truck was in October 2016, when two thousand cases of beer were driven 120 miles within Colorado by a truck from Otto, the Uber-owned vehicle company. When I read about these trucks, I thought of Steven Spielberg's early film *The Sugarland Express*, with programmed yet suddenly out-of-control vehicles barreling murderously down highways, their grills forming into nasty smiles. But this scenario is not what I worry about. Should trucking jobs go away for millions of workers, we're likely to see more people out of work and trying, often unsuccessfully, for a second act: more families scraping to pay for rent or day care, and yes, perhaps more voters enraged enough to pull the lever for demagogic candidates who stoke their anxieties.

Some activists are so concerned about automation in trucking and about driverless vehicles that they're already starting to organize. The not-for-profit organization New York Communities for Change (NYCC), for instance, has been agitating against automation in trucking and driving and launched a campaign targeting the U.S. Department of Transportation, which has billions of dollars set aside to subsidize the development and spread of autonomous vehicles.

"Many truckers are very fearful," said Zachary Lerner, the group's senior director of labor organizing, who has been organizing drivers against the autonomous vehicles. "Trucking is not the best job, but it pays the most in lots of rural communities. They worry: Are they going to support their families? And what will happen to all of the small towns built off the trucking economy?" Driverless Ubers may indeed threaten the gig economy freelancers we met earlier—the schoolteachers who drive for rideshare services in order to pay their bills. (Ironies compound: as the writer Douglas Rushkoff has noted, today's drivers are themselves now part of the research and development for what will most likely be the driverless future, building up a company with their labor in preparation for a time when the company will do away with them.)

"Our demand is to freeze all the subsidies for the research on autonomous vehicles until there is a plan for workers who are going to lose their jobs," Lerner said. As part of this effort, NYCC regularly puts together conference calls between dozens of taxi, Uber, and Lyft drivers. They discuss how they've all gotten massive loans to buy cars for Uber and how they are still going to be paying off these loans when the robots come for their jobs—the robot vehicles Uber has promised within the decade.

The robot-fearing Middle Precariat also includes parts of the legal profession: robots are threatening higher-end jobs, including those usually carried out by humans handling information. I first discovered why lawyers were worried in 2015, when I went to LegalTech, a law and technology conference where thousands of attendees at New York City's Hilton Hotel checked out various law-related projects—most prominently, software that reviews legal documents. There was the prerequisite small plastic gavel swag and the former corporate lawyer building a large-scale Lego version of Van Gogh's *Starry Night,* but there were also lots of booths advertising software that basically helped firms reduce the number of their employees, including attorneys. Such technology might soon make some of the lawyers on that trade-show floor extinct. Indeed, a representative of one of the software companies told me on the trade-show floor that the greatest thing about their product was that it needed barely any workers to run it. As we already know from two of the previous chapters, lawyers are increasingly under- and unemployed, in part because of the robotization of law.

The impact of technology on the legal profession is a big deal, as lawsuits usually involve thousands of documents that lawyers and paralegals are paid by the hour to review. The review has long been the dreary bottom rung of any legal case, with legal worker ants sifting through every piece of evidence to see if it fits. While it's often nasty, detailed labor, nowadays a job reviewing documents may be all that a law graduate can get.

Even if you can get such work, the proliferation of legal technologies squeezes law workers and drives down their pay. A former staff attorney at one document review mill—who bitterly called the employees "doc monkeys"—described Dickensian scenes of workers being bussed deep into Ohio or Pennsylvania. She believed that "robotic" software services had lowered the document review wages for humans so that all that's left is work in these kinds of precarious and underpaid sites.

"Doc monkeys" are typically now earning just $17 to $20 an hour, while shouldering upward of $200,000 in student debt. They usually have law degrees. In a sample ad for such a job, a company called Business Intelligence Associates (BIA) offered $20 an hour to both recent law school grads and licensed attorneys for temporary work doing document review, trial prep, and support work. The ad claimed that the perks included a "great work-life balance."

However undesirable, jobs like these may now be the only ones available to recent law school graduates. When e-discovery systems, like the ones on display at a legal tech fair, get more sophisticated, even these jobs will be gone, but first they will linger on for a few years, while lawyers are paid less and less.

Sometimes those squeezed by robots are nurses like Bonnie Castillo, associate executive director at the California Nurses Association. Castillo had been in the profession for many years, including a position in an intensive care unit. "We've been very concerned about robots for years," she said. When she first started to hear about robots coming to hospitals near her, she felt instantly degraded and, as she puts it, "commodified." Nurses, she said, are on a "conveyor belt, controlled by the health-care industry." (Elsewhere around the country—in hospitals in Saint Paul, for instance—nurses have protested against the use of algorithms to run their hospital floor. The nursing union thought that the use of algorithms was deeply unwise: why remove human judgment, which factors in family and psychological issues and which robots are incapable of, from decisions regarding patients and staffing?)

In their work, Castillo and other nurses discover things about patients that she didn't believe a robot could discern because patients require a human touch, as she put it. She meant that quite literally. As a nurse, she said, she would typically assess the texture of patients' skin, determining, for instance, whether their skin was clammy or sweaty. Would a robot dispensing medication note that a patient was taking a turn for the worse through such subtle signs? Or would it just roll by after completing the delivery, having missed the opportunity for a patient contact that might have led to a life being saved or prolonged? Castillo said: "A caregiver-patient relationship is essential to nursing." That connection has to be established in minutes for patients to get the help they need. "People put their lives in our care: are robots empathetic?" Castillo objected to the robotization of a "female-dominated profession" like nursing, especially after the recession.

If Castillo was unhappy with her robotic colleagues, so was Mardi Thompson, a nurse in the neonatal intensive care unit at University of California, San Francisco Medical Center at Mission Bay, who told PRI's *Marketplace* that she was concerned about her profession. People "need jobs," she said. "And then we have the robots doing them."

As Castillo put it, nurses "are the heads of households and are often supporting whole extended families." Nursing robots also threaten a profession that actually encourages mobility and is a growth industry. The Bureau of Labor Statistics projects that the demand for nurses will grow by 16 percent from 2014 to 2024 as the elderly compose a bigger and bigger proportion of the U.S. population. It is a classic route for many—women and immigrants in particular—into the bourgeoisie.

And of course, I didn't have to look far to find Middle Precariat journalists at risk of being replaced. They were usually the recipients of grants at my nonprofit, the Economic Hardship Reporting Project. They were up against the new trend at Tronc, formerly known as the illustrious Tribune newspaper company, which in its inadvertently hilarious branding videos celebrated artificial intelligence over that of photo editors, reporters, and the like, replacing them with optimization and something called "content funnels." Tronc's representatives used terms like "machine learning" and "artificial intelligence" and claimed to be able to automate tasks, like searching for photos, that were once the province of photo editors or photo assistants. These services are automating the production process to make journalism cheaper and, yes, worse. The full terror of it came home to me after Donald Trump was elected president and began tarring journalists. Will software be able to stand up to him or any other bullying financier, especially one who denies facts?

Meanwhile, sites like Automated Insights, whose very name is an oxymoron, use algorithms to generate stories in publications such as *Forbes*, generating one story every thirty seconds in this fashion. This process may replace writers like me who freelance for publications like *Forbes*. The Associated Press regularly publishes stories of companies' quarterly earnings—"Apple Tops Street 1Q Forecasts"—without a byline, because they are written by a computerized system that has memorized the *AP Stylebook* rather than by a flesh-and-blood reporter. (Every quarter, the AP publishes three thousand "robot"-written stories with Automated Insights, which has the unsubtle and frightening acronym AI.) The incentive for the AP is not only to save on labor costs but also to write up business news before anyone else can—literally, before a human being could—and with no misspellings. But software—surprise, surprise—is a flat storyteller. The robot "voice" is formulaic and predictably impersonal. An automaton would not get into the Iowa Writers' Workshop (insert snarky joke here). It can't analyze and isn't even as good as the worst j-school students at getting a decent quote from a source (even, one would imagine, if the sources were other robots). The writing lacks specifics and the level of precision that even a mediocre reporter would be able to muster. A robot can't spot telling details about people or events, nor can it organize information in a compelling fashion. Automated Insights reads to me like the worst reporting from any journalism school. As a student at journalism school and also at times as a professor, I feared having to use or teach the pyramid structure: the newspaper article outline for creating news stories—to which traditional newsrooms slavishly adhere—like a player piano "creates" music. Yet the software has embraced such rote and dead-eyed aspects of my profession. The question is, do readers of this automated content care about its lack of tone and specificity? Do they even notice that it's inhuman?

Another concern of robot-doubters like me is about what happens if the math of robots doesn't really work out for humans. Given that, over the course of ten years, what it costs to "pay" for the robot or pay a human employee might not be that different—the robot might even be more expensive—and given that middle-class jobs are hard to find, their numbers dwindling, why would we want them to fill those jobs that still exist? The money saved by *not* having to employ a human being is supposed to make up for what it costs to buy the robot, but it doesn't always. For example, a robot pharmacist can cost $15 million: does that actually save money for hospitals? At $100,000 a year, a human pharmacist would have to work 150 years to earn as much as what her robotic successor costs up front. The argument that these robots will save money over the long term relies on an absurdly long term. The theorist Zeynep Tufekci agrees with this second position, writing that robot caregivers are "economically destructive" and that accepting them rests "on accepting current economic policies and realities as if they were immutable."

I wondered: what if the critics of robots, like the labor organizers I spoke with who are trying to get truckers to resist driverless trucks, could explain their position to robot-lovers and techno-positivists? Or what if we simply decelerated our robot interlopers and established a "slow tech" movement to match our "slow food" and "slow fashion" trends? Or at the very least, what if we started to rethink who *owns*, say, autonomous trucks? The effect of robotization would be profoundly different if truckers themselves owned their own autonomous vehicles rather than a corporation controlling them all. Robots are always spoken of as emblems of the future, but they seem to me part of the past as well, akin to the Luddites' weaving machines. What if the Luddites had been members of co-ops, with a stake in the automated looms that replaced them?

IF I HAVE GIVEN SHORT SHRIFT TO THE ROBOT ENTHUSIASTS, IT'S due to my personal predilection for people over efficiencies.

The "robot-chasers" tend to gloss over the impact of losing human jobs to our mechanical colleagues. They focus on the efficiency of hospital robots, for instance: how autonomous they are, how little system interaction they need, supposedly, to function. If a TUG meets a problem it cannot resolve on its own—perhaps it is asked to move a bed and the bed is for some reason not where it is usually found—it connects to a "cloud command center," whose "people" solve the problem most of the time. Robot enthusiasm can be heard in the gleeful descriptions of Japanese medical robots, creatures like the three hundred-pound Robear that gently carries patients to and fro. It can be sensed in the chipper 2014 *New York Times* op-ed by a geriatric specialist insisting that now is the time for robot caregivers because, in a world with too few carers, we need these mechanical helpers to handle the sheer number of older patients. And it's in the lingo of robot manufacturers like Aethon that praise their creations for their "24/7 Improved Productivity" (a mechanical twist on "the forever clock" that is now being imposed on human workers). "TUG works around the clock," Aethon's site chirps. "It is a substitute for the labor needed to haul and transport goods, materials and clinical supplies in the hospital."

Robot fans may praise their self-sufficiency and their inhuman discipline. In the latest media accounts, you often find words like "efficient," "polite," "uncomplaining," and "cute." The use of the word "cute" here seems evocative, as if the humans are trying to accept, diminish, or make peace with these animated objects that may well in the end displace them. As the theorist Sianne Ngai puts it, cuteness is a "sentimental attitude toward the diminutive . . . [objects that are] formally simple or noncomplex, and deeply associated with the infantile, the feminine, and the unthreatening." In articles, robots are commended for not getting tired or needing a break, no matter how many hours they work. At Sinai Hospital in Baltimore, they have cute first names: Rigby, Herbie the Love TUG, Jake, and Elwood. At Presby, they also have pet names: R2D2, C3PO. At that hospital's fair, the robots spoke to the hospital staff, gave out candy, and dressed as a pirate and a pirate ship. When I heard about this fair, I thought: *These are scabs in robot clothing!* I have been told that pharmacist robots make fewer mistakes than their human counterparts. At the Massachusetts Institute of Technology, the Nao, designed to assist nurses in a labor ward, nudges doctors on where to move a patient and which nurse should be helping with a C-section, and it is supposed to "fix" scheduling in labor wards. MIT's Julie Shah and her coauthors, in a paper on the Nao, present it as taking a real burden off of workers. Similarly, according to L'Altrelli, the TUGs at Pittsburgh's Presby are the outcome of "restructuring" at the hospital.

The robot-positive camp sometimes also asserts that cyborgs are the key to liberating workers from boring and even abject tasks—mind-numbing and alienating tasks that (paradoxically) dehumanize workers. What if we could be freed of such work? The argument is that schlepping clean linens or removing trays isn't "meaningful" work for the nurses and janitorial staffs who have long performed these exhausting and rote tasks. (I don't necessarily agree with this argument, as at least some of this work is care work, which I consider the most crucial work of all.) And isn't trucking often miserable, what with the overwork, the speeding, the risk of accidents, the sheer banality of driving thousands of miles, being far away from home and family, living in motels, and becoming obese eating fast food? And isn't it important for humans to find meaning in their work and thus in their lives?

The third framework, which sees both the downside and

the upside of robot workers but is simply more relaxed about the whole issue, is perhaps the best of all. It's roughly the sentiment held by people like Martin Ford, author of the 2015 book *Rise of the Robots,* who thinks that automation is a problem, though an inevitable one. In an interview, he told me brightly that he is a futurist. This third camp holds that robots are not such a big deal, but that we should also try to do something about their incursion. As Ford put it to me, we will have to learn how to address any underemployment that results from their mechanical rivalry.

And then there is the fourth camp that I mentioned earlier—they definitely fear the march of the robots, but think that everything may be okay if we embrace universal basic income, or UBI (or the more catchy-sounding BIG, for basic income guarantee). This fourth group includes people like UBI "ambassador" Scott Santens, an author and advocate who often writes in support of these initiatives. One reason why UBI is so necessary, Santens told me with an enthusiasm that veered between that of a zealot and a bubbly partygoer, is to protect us from the inevitability of a robot workforce. True to his geek identity, he said that he was thrilled by the thought of a robot caregiver. He wanted one himself in old age. (He was still only forty.)

Santens's inspiration for his work lay in his own biography. He himself had been self-employed his whole life, designing websites and the like. "I never had a safety net. I've never experienced health insurance paid for, or a 401(k), or insurance. None of it has existed for me." He lived in New Orleans and raised his income through crowdfunding, which he saw as a kind of primitive UBI—250 people support him each month. When I asked him how he thought of himself, he quoted Martin Luther King as a basic income supporter.

According to Santens, universal basic income would be an annual stipend of a set amount, distributed by the government to each citizen, whether they worked or not. In the United States, UBI would most likely be set at the poverty line: individuals, for instance, would receive $12,486, the poverty threshold for individuals, and a household of three would receive $19,318.

Although this idea has been around for a while, it's gaining new traction with a broad array of people, from the left to technology "ambassadors" to conservatives—all for different reasons but with the same results.

"The amount of unpaid care work that is going on right now is estimated worth $700 billion a year," Santens told me, up from $691 billion in 2012, and roughly 4.3 percent of the U.S. GDP. That number includes day care for children, but also adult children taking care of their parents and older couples taking care of each other. Santens fantasizes that UBI could become paid maternity leave for moms of newborns as well as replace many benefits. Proponents like Santens think that UBI could help make sense of the automation of so many middle-class and working-class jobs. It would protect workers who lose their jobs to automation and thus alleviate the impulse to blame themselves or, even worse, point fingers at immigrants and people living below the poverty line.

As for how we would pay for UBI, advocates insist that it is not as expensive as it might appear. We could raise money with a flat tax. And UBI could partly or fully replace existing safety net programs, such as Medicaid and Social Security. Further, it could help eliminate some of the hidden costs of poverty—like the medical bills both the insured and uninsured incur.

While no country has yet nationalized UBI, it is more

than just a fantasy. In the summer of 2017, Ontario, Canada, piloted a UBI project with four thousand participants. The pilot offered a minimum baseline of money calculated by income—whatever people earned over and above this amount—and they continued to receive benefits. The *Ottawa Citizen* kvelled about the project as another example of how their country was coming together while the rest of the world was falling apart: "The virtues of a universal basic income are, well, universal. It grants people a financial cushion to deal with things such as sudden job loss," or a dangerous pregnancy. The Finns are also implementing UBI with an admittedly tiny sample group, and UBI has been piloted in areas of India and Namibia.

UBI has had a complex history in America. Richard Nixon argued for it in 1969. In an experiment in the 1970s referred to as "income maintenance," citizens in certain cities, like Gary, Indiana, and Seattle, Washington, were given a certain amount; that experiment ended in 1982. Since that year, Alaska has paid residents something resembling UBI—a state dividend that on average offers a small set amount a year in additional income to men, women, and children, usually $1,000 to $1,500 each.

Today, UBI's supporters are a wild mishmash of types: venture capitalists, Democrats, and Charles Murray from the right. The feminist theorist Kathi Weeks thinks of basic income as not just a leftist or giddily disruptive answer, but rather as a fine response to our nation's failure to financially support, or even notice, female labor in the home, from housework to child care. Automation has gotten so extreme that work, as the Rutgers University historian James Livingston writes in his smart and rude book *No More Work*, is "no longer socially necessary, which means it doesn't pay. The labor market is broken, and it can't be fixed. Or it's been *perfected* at this outer edge of late capitalism." It is getting easier and easier to use capital to replace human labor—to substitute machines for "real, live human beings" to the point where capital and labor start to seem equal to each other. Livingston is against work, or at least the traditional work that he considers close to wage slavery. After all, wages haven't kept up with the cost of living or the expense of education, and we have lost our race with machines, as he put it to me. If the societal need for work is gone, only the individual need for work's income will be left, he writes. Then labor will no longer be how we should define ourselves. It will be less likely to be the source of a decent living, and no longer a way to measure our strength or our virtue.

After speaking to the UBI advocates and paging through their various polemics, I started to imagine how my parenting experience would have gone if it had been underwritten by UBI. I imagined an alternative or double life, a choose-your-own-adventure with a different outcome than my real one. If UBI had existed, perhaps I would not have needed to constantly leave my daughter's side to check my email for potential work leads. Would I have been checking my dwindling bank account while I screwed the pale yellow plastic tops onto her bottles, doing crucial arithmetic in my head? What if I hadn't been totting up all the months of pay I had lost, but rather had in a sense paid myself for caring for my daughter out of a basic income guarantee?

After all, according to its cheerleaders, UBI would offer modest monetary support for *any* unpaid carer—a parent caring for a child, a wife caring for a spouse, or a grown child caring for an elder. What if UBI could help with both the problem of the end of the middle class and the problem of the devaluation of care work—partly or even mostly because it is seen as feminine and thus looked at through the prism of sexism?

UBI might be of particular value to mothers caring for children and caregivers for elderly parents for all the difficult work they do free of charge. As the journalist Judith Shulevitz wrote in the *New York Times:* "The U.B.I. would also edge us toward a more gender-equal world. The extra cash would make it easier for a dad to become the primary caregiver if he wanted to. A mom with a job could write checks for child care and keep her earnings, too." In other words, UBI might give people who care for their loved ones for free a cushion that would economically and socially support their choice to nurture. That could well lead to a reframing of how we think of care overall. If UBI reigned, affection would be not just a sentiment but an ethical practice and also a form of accepted work: it would be legitimate that we could be paid to love. UBI, as Santens and others argue, would allow humans to focus on the child care that has until now stretched families to the limit.

Back when my daughter attempted to breast-feed in the manner of either an inconsolable alcoholic or a small (and very pretty) pig rooting for truffles, back when I devoted myself to understanding her basic needs, she liked the noise of showers with the lights off, her father's voice with its soft radio lilt, bumblebee pictures, being cleaned with olive oil and lanolin and cotton balls, the word "world," the phrase "beautiful girl." She liked being over a shoulder, spying on an apartment because an apartment was the whole world to her. This "reading" of a nonverbal creature was all-consuming and exhausting labor. Caring for her was as full-time as it could get. And yet I needed to *actually* work too. Ironically, I even depended on (low-level) automation to help me so that I could edit while my daughter relaxed: her Fisher-Price mechanical swing rocked her back and forth so many times faster and for so much longer than I could. (Plus, it was fluffy and adorable, built to resemble a sheep.) As an infant, she often fell asleep this way, in a vibrating robot sheep's mechanical embrace. I knew I was supposed to be rocking her in my orange nursing chair, swaddled in a white cotton blanket with ducks printed on it, to satisfy her desire for repetitive movement, but the auto-rocker seemed to be doing it better. Luckily, I was still in the same room with her, although concentrating on my laptop, which to this day she still considers her sibling rival.

Would UBI have helped me back then, so that I wouldn't have needed to take temporary editing jobs? And more important, might it help all mothers who care for their children out of far more fragile positions than my own?

THE IDEA THAT AMERICA SEEMS *NOT TO CARE ABOUT CARE*—AND TO deprive caregivers of money and respect—emerges again and again in *Squeezed*. It helps explain why middle-class parents can't afford life in America, and it underlies the unchecked rise of the robots as well. Anyone who doesn't care about care cannot see the informal value of, say, patients interacting with medical staff after surgery, even when they are only picking up laundry or taking away strawberry Jell-O containers. I thought of the nurses who came by to see me when a complication kept me in the hospital for a few extra days after I gave birth. These human exchanges were so necessary as I stumbled, on massively epidural-swollen legs, in a robin's-egg-blue hospital gown, surfacing for the prerequisite photo of me holding my daughter, her face frozen with the shock of being born, as beautiful and thin-eyebrowed as a silent film star. My body, under a black silk robe, was in total disarray. But those brief interactions with nurses made me feel that much more human. I remember their faces through the haze of pain or as I hobbled down the hall. I remember their consoling words. Such moments could well be lost in the future.

Few patients would argue against rapid or accurate service, yet I also experienced distaste at the thought of a hospital robot, both as a daughter and as a mother. My elderly mom had some serious surgeries in recent years—both knees replaced and more—so when I asked her about the TUGs, she easily imagined herself in a hospital bed with laconic robots doing even small things for her, like taking her linens away. The thought that robots would take care of her "when you are most vulnerable," as she put it, seemed deeply wrong. She felt as I had, with the nurses after I had just given birth. "You need the feelings of a real person around you," my mother said.

Indeed, instead of rewarding hospital workers for the small moment of humanity they share when they bring a patient a lunch tray, we overlook, demean, and poorly compensate that labor. If automation forces us eventually to uncouple income from work—to recognize that robots have truly replaced us and as a result we need to pay people for not working—we will have to cease making moral arguments, from both right and left, about the value of work. But even more, we'll need to think differently. We'll need to learn to value various kinds of nonwork behavior that we do not honor now—and also to protect and value the kinds of labor that incorporate these nonwork behaviors. What happens to love after work becomes obsolete? And especially, what happens to care work, the labor that most closely resembles affection? Shouldn't there be at least one sort of labor—care work—that must not be mechanized? How we regard the work of nurses highlights how we demean care. If we valued care, we might object to efforts to replace human beings with robots at the bedsides of suffering patients or mothers giving birth.

But more broadly, instead of giving robots personhood status, as was being discussed at the time of this book's writing, we should concentrate on protecting our human workers and, by association, their families.

Even those who may lose their jobs thanks to robots tend to seem unable to muster hatred for their future mechanical rivals. As Finn Murphy, the author–truck driver, told me fatalistically: "I am not going to take the Luddite perspective—driverless vehicles are going to happen. They can put wrenches in the weaving machines and there will still be these trucks." He seemed to just want to be seen as a realist. He certainly didn't want to be caught out as a deluded idealist, pushing back futilely against progress.

Murphy wasn't the only one. I heard such sentiments often, and I couldn't help but sense in these sentiments an acquiescence that was, to my mind, symptomatic.

For starters, why shouldn't we be Luddites? And if not Luddites, why are we—and truckers—not advocating for the cooperative approach to ownership of autonomous trucks so that millions of drivers are not left out in the literal cold? We might say that, if robots prevail, *most vocations* will actually be the "impossible professions" Sigmund Freud wrote of—not just because they produce "unsatisfying results" for the worker but because they barely exist for humans at all. Save for *the profession of owning robots*, that is.

For many other job categories, perhaps soon to be rendered impossible, from nurses and legal assistants to movie ushers and cashiers, shouldn't we be concocting legislation to help all strata of workers who will be displaced by our mechanical friends? And we should certainly reintroduce care work into our discussions of automation as the most important kind of labor that should remain robot-free, as the most vulnerable humans need to interface with other humans. We might stand up for the values of humanity more directly, at least asking why anything that throws millions more human beings out of work is considered "progress."

In the future, perhaps all of us non-robot-owners may have to eke out our income from whatever remains. Centuries ago, in his story "The Automata," the German Romantic writer E. T. A. Hoffmann wrote: "Yet the coldest and most unfeeling executant will always be far in advance of the most perfect machines." The warmth and feeling of human beings must be honored, at the very least. If we don't at least try to make the future more equitable, most of us will be left with scraps.

CONCLUSION: THE SECRET LIFE OF INEQUALITY

I started this book in mild personal distress. My life and those of the people around me had become tenuous after the Great Recession. My friends worked as permanent adjunct professors or journalists who didn't make enough per word to support their children. My broader circle included a photo editor, administrators, and a carpenter, who were all out of work.

The friends and acquaintances sought advice when I started directing and editing a reporting nonprofit dedicated to journalism about inequality. I was developing articles and essays on hardship and financial consternation. I assigned and funded pieces by the formerly middle-class journalist with three kids who was living without heat, tapping out his stories on his phone; the author selling his plasma just to get by; the strong single mom who had worked as a maid and had been on food stamps but had somehow never stopped writing. I was hearing firsthand how the recession was having a much longer half-life than common wisdom had it.

My first bit of wisdom for anyone who asked was: don't

blame yourself. Self-blame seemed omnipresent and corrosive: it oozed through my friends' voices and into my phone's headset after my daughter fell asleep; it was a murmuring beat under conversations between parents at parks or playgrounds. *You are doing the best you can for your daughters,* I'd tell the mom castigating herself about her child's school. A change in school would have required the family to rent an apartment in another neighborhood. And this strategy seemed politically suspect as well: they'd be moving there simply to go to a public school filled with wealthy people who raised money only for their own school. *You just can't do better,* I'd want to say. *There are larger reasons why your job is precarious and your parents' jobs weren't. You can't help that you are being evicted from your apartment to make way for luxury condos. It's a system failure. It's bigger than you.* At some point my advice started to sound like an inner monologue, recited as much to still my own mind as to help those around me. *You can't help that the public schools in your district are failing, that you can't find common ground with your neighbors to create shared, reasonably priced day care, that you wound up seeking scholarships at a private school and left the commonweal entirely.*

I'd stab my iced coffee with a straw, standing in the urban sunshine as I offered these nostrums. My sleep was also full of advice, seemingly to myself from my troubled unconscious: I had nightmares about housing shortages and school choice.

Initially, I feared for my family, as for a while neither my husband nor I had any work except freelance writing. While things turned out just fine for us in the end, I knew statistically as well as firsthand that workingwomen had it worse. Always worse. The mothers I knew and the ones I gave grants to through the nonprofit tended to make less than the males around them. And for the more privileged, the green scales of optimism had simply fallen from their eyes. For those who were members of groups who have been historically excluded and oppressed, from single mothers to parents of color, their financial squeeze had sometimes become a strangle.

Leaning in and *work-life balance* very quickly seemed like hooey. Despite their education and training, some of these women were just trying to get by. As the working-class feminist icon Dolly Parton has said in response to being questioned about Sheryl Sandberg's book on corporate feminism *Lean In:* "I've leaned over. I've leaned forward. I don't know what 'leaned in' is." When I read Parton's remark in an article I edited by the writer Sarah Smarsh, I thought of the adjunct professor moms on food stamps, the lawyers consumed with resentment as they tottered under their student debt bills, and the workers experiencing pregnancy discrimination—all of them were *leaning over.* The caregivers and parents knotted together into twenty-four-hour day cares were *leaning forward.* So were the schoolteachers and former journalists forced to drive for Uber, as well as the mom I spoke to who got her kids enrolled in voluntary but troubling medical testing for extra cash, including having MRIs done of their brains. I am thinking also of John Koopman, the dad who used to wield a steno pad and pen when he was a war reporter. After layoffs, he managed a strip club, where he broke up drug-fueled fistfights and carried blacked-out dancers rather than recording equipment.

One mom said laughingly that her family economic strategy was having an only child. "Being relatively infertile, we haven't gotten pregnant again," she said. This quip was not just a one-liner. As Lauren Sandler writes in *One and Only,* professional women, myself included, are having fewer children to get by.

I saw that the future for these parent-workers was likely to grow even dimmer. Despite calming murmurs from some economists as they chilled out over the rise of the robots at their think tanks, opining that the automation revolution has been oversold, I read and spoke to enough people to see that, unless we prepare ourselves, a robotic hell is indeed awaiting us as workers. (With the phrase "robotic hell" I flash to the film *Barbarella*'s army of mobile dolls attack.) Robots are expected to badly impact working- and middle-class professions. The jobs potentially to be replaced run the gamut from butcher to pharmacist, tax preparer to taxi driver. As Stanford University professor Jerry Kaplan has written, "the color of your collar" doesn't matter to these mechanical creatures. A 2015 McKinsey study found that roughly 30 percent of the tasks within 60 percent of our current American occupations could soon be turned over to robots.

How might my just-middle-class friends and more financially fragile writers or sources find relief? The battle will be for survival, but it's also against shame. Jeans-clad, juice-box-wielding, we are trying to find our way out of the maze of our country's idea of parenthood.

WHAT IS TO BE DONE?

I started tracking solutions to that which plagued the parents. Promising responses include small-scale debt consolidation, coparenting arrangements, student debt forgiveness, and adequate workplace protections for pregnant workers. (You have read about most of them and many others.)

I learned again that with better, cheaper, and more accessible day care, women would be much more likely to be employed and families much less likely to be unstable. I discovered again that most of our political and greedy corporate pashas are evading these issues: they have pretended that parenting and family are private matters and solely our individual responsibility.

One fix would be a universal child allowance scheme in which all Americans would receive additional money simply for having children. Our tax system supports some families with children through the Child Tax Credit (CTC), which can be worth $1,000 per child, and also through a child tax exemption. But there is increasing discussion of making it more substantial. With an allowance, each child-rearing family would receive cash grants. Internationally, such a grant is the norm: in Sweden, for instance, the kid allowance is considered money for rent, smorgasbords, or additional baby furniture. How barbaric that in our country many of us get so little support for raising our children.

Beyond a broad allowance for families, the elemental solution would be high-quality, across-the-board, way-better-subsidized day care. For families headed by one parent or where parents both work, the need for a day-care *system* is clear—something akin in philosophy and reach to the much-embattled Obamacare—for the adequate nurturance of children across our land. That decent child care is currently a sign of social privilege is simply outrageous. Broad-based, accessible, adequately subsidized day care would also make the lives of day-care workers like those you've met in *Squeezed* more secure as well; the women earning their livings in the child-care field are usually subsisting in dedicated poverty. In the existence of said workers, we can see how class, gender, and race intersect, parents and caregivers in a troubling economic symbiosis, like Matryoshka stacking wood dolls. As Elise Gould noted in a report from EPI in 2015, 95.6 percent of child-care workers are female. An out-of-proportion number of this estimated 1.2 million workers are immigrants and people of color and one in seven live below the poverty line.

But how would an elaborate, humane, and expensive day-care system that could correct our current one ever come to pass in this country? As an answer, some not-for-profits and a company have banded together to launch the Who Cares Coalition to create a "social change movement" binding together families and caregivers, a marketplace named Care.com, and a think tank replete with a trendy neologism to describe the workers for whom it advocates: the careforce. The revolution they hope to spearhead will, of course, be hard to carry off in America, where we are squeezed on all sides, but these advocates are making some headway, including the formation of a growing number of nannying and domestic workers' cooperatives.

The more likely improvement, at least within the span of my own daughter's childhood, is widening access to something we've already discussed: universal public pre-K. I first encountered the program in our city, New York. To understand universal pre-K better and discuss ideas for nationalizing it, I spoke to Richard Buery, coordinator of the city's universal pre-K program under Mayor Bill de Blasio. How had the de Blasio administration done it, and how could such a program spread further faster?

The rollout of universal public pre-K was surprisingly fast, ramping up from 20,000 pre-K seats in January 2014 to 51,500 in September 2014, to 68,000 seats in September 2015 (though they had planned for 73,000 by then). Buery said that the secret to getting universal public pre-K running—beyond the $1.5 billion five-year state allocation funding it—was cross-city agency coordination: more than twenty city agencies were involved in the effort. "It was its own mini–military operation," Buery said of what made up the majority of his job at the time. That's lesson number one for applying New York's universal public pre-K nationally: organize across agencies, both federally and locally, as quickly as possible.

What made it work in New York City, Buery said, was using existing infrastructure—Catholic or Jewish schools, for example—as additional pre-K buildings. Indeed, the city found thousands of educational spaces and recruited thousands of teachers.

That's lesson number two: step past the fictional barriers thrown up by precedent (for example, "Oh, we can't use religious schools as sites for a government program"—sure you can, and you should) and streamline ramp-up times and procedures, slashing red tape. I kept thinking while Buery spoke that this could—and must—be executed in different places around the country. If a place as huge and wildly diverse as New York City could roll out universal pre-K, it surely could be brought to smaller places. The main principle here was not to be too finicky. Buery's advice to other cities for nationalizing universal pre-K—besides having the public and political will and the coordination—was to use school resources already in place, no matter how downbeat or unusual. "Every community has certain assets—every place needs to start where they are."

A third lesson: establish a vigorous feedback loop involving the served population. By organizing a system of focus groups to solicit feedback from klatches of parents, New York City legislators learned new things—for instance, according to Buery, respondents said that they saved an average of $10,000 to $15,000 a year thanks to public prekindergarten. This much-needed cash would have otherwise gone to private preschool, day care, or a babysitter, or it would have come out of the working hours and pay of parents who would have had to stay home and provide care. But it was the psychic burden more than the financial burden that seemed to lift, said Buery, not just for poor families but also for middle-class families. "You can make $100,000 in New York and still have trouble making ends meet," added Buery.

The city is now developing another, possibly replicable program called 3-K for All. ("I thought it was a clever pun," Buery said.) 3-K will serve the city's three-year-olds, starting with two low-income school districts that include Brownsville and the South Bronx.

A third potential, and admittedly huge, fix is something we discussed in the last chapter: universal basic income. In 2017, Ontario, Canada, began implementing a three-year pilot basic income guarantee (BIG) program. In three sites in the province, Ontario is offering individuals a set amount of money per month, up to $24,000 a year per person. In giving people a living wage, it is expected to cost $50 million annually and is open to any single person earning less than $33,978 and couples earning less than $48,054.

When I spoke to Canadian BIG advocate Roderick Benns, he explained it as a kind of societal bolster pillow that will not only allow parents to care for their own children but also challenge "the ideas that paid labor is the absolute definition of what it means to be human." Benns was poetic, even ruminative: "Care for our children is work, after all, as is the administration of one's life."

UBI and BIG sound wacky, as tech au courant as the ridiculous Silicon Valley business appellation "chief evangelist." But some U.S. states are now taking the idea seriously: Hawaii is looking into UBI to help its working families deal with the insanely high prices of their island paradise.

Of course, grand ideas like UBI require more than an army of techy disruptors pontificating at TED conferences.

Establishing UBI or BIG programs requires political will and amenable leadership. To get that, we need a new coterie of state and city officials who will fight for paid leaves, a progressive tax code, and, yes, policies like UBI. We will need to throw our weight behind progressive political outsiders, which is a whole other book for someone ambitious enough to write.

A fourth big shift would be for corporations to address their employees' day-care needs. This would benefit so many Americans. When have companies actually shown understanding for parents or the caregivers of their employees' children? Very rarely. Some companies do have on-site "babies at work" programs; the roughly two hundred such programs found in the United States represent a small proportion of all our corporations, according to the Parenting in the Workplace Institute. These companies include the baby clothes company Zutano. (I immediately ordered a pair of Zutano red bike shorts for my daughter upon hearing of this.) Zutano has had this arrangement in place for more than a decade, and its founder has said that participating parents are the most long-standing and "loyal" of employees. Let that be a lesson to larger companies seeking low employee turnover.

Other small corrections include ratings of "corporate culture compliance" that applaud the best companies and shame the worst with respect to how they treat their parent-employees. Companies that do not want to be baddies have built family needs into their business models. A report issued by the Center for WorkLife Law cited fifty such organizations, all in the legal field. These companies include everything from virtual law firms to "full-time flex" policies where workers can work at least partially from home.

All of these fixes, however, are only beginnings. They are sketches and mini-blueprints.

What to do while we wait—probably much longer than our childbearing years—for an improved social architecture?

SOLUTIONS OF OUR OWN

We can try to change how we feel and think about our own benighted condition. I hate to suggest that this alteration must come from within, as if I were a yoga teacher forcing a student into an even more relaxed corpse pose. But how else should we deal with our leaders' severe neglect?

Coparenting and other DIY collaborative day-care solutions require sharing with others and repressing your personal agendas in favor of the greater good. The self-control these boutique solutions sometimes require is beyond Buddhist, and I am loath to recommend them as comprehensive solutions to the feeling that the proverbial deck is stacked against you. Some artisanal fixes, after all, rest on quietism—the practice of retreating from a dreadful world into the self, mini-networks, and small, atavistic activities and pleasures. I like to think of such internal shifts as all of us creating a sort of shadow state, whether we band together with small swarms of like-minded people or "work on ourselves," all in order to tide us over while we ride out the current wave of government failure.

The first internal shift is this: *stop placing blame*. Parents either blame themselves or blame others for their problems. We need to change both of these reactions.

On the first count: those who blame themselves believe that, say, being unable to afford their health insurance is their failing alone rather than a system failure. And those who blame others believe that they are "cutting the line," as the sociologist Arlie Hochschild has framed it. Kathy Cramer, a political scientist at the University of Wisconsin who studied her neighbors in her state, found that they believed they'd gotten the short end of a virtual stick. They imagined that their "rural values," their tastes and beliefs, had been overlooked or cast aside and that other people—urban elites, immigrants—had pushed in front of them.

Yet self-blame and other-blame are two sides of the same coin. If "self-blamers" could gain greater critical distance, they might begin to see their problems as part of a societal snafu. They might then see themselves as less culpable and realize that they have more in common with their friends and neighbors than they might have thought before.

Second, we need to *reframe care*. Of course, we need to acknowledge first that we are up against a harsh social reality around caregiving. As I just mentioned, caregivers are underpaid and undervalued, while corporations make little time and space and offer even less money for their employees' child care or maternity needs. It's no coincidence that many of the people you have met in this book are care workers: nannies, day-care directors, teachers, professors, and nurses. All such work, I'd argue, is especially poorly paid because *care work is devalued*—even *disdained*. There is a stigma around any care-related profession. Take school teaching: Americans and their representatives give lip service to teaching yet continue to ruinously underpay teachers, and now they are attacking teachers' unions.

But on our own, we *can* begin to think about care differently. For instance, we can support care worker cooperatives and unions, both in real life and online, like the ones I write about in this book. We can hire from worker-owned cooperatives, like the New York City housecleaning service Brightly Cleaning, rather than from cleaning conglomerates like Handy. We can also reframe care out of respect for the domestic workers who keep so many of our families together, as Ai-jen Poo, who has organized domestic and care workers for years, argues. We can also ask ourselves more philosophical questions and challenge our assumptions. Why do we identify care with weakness? Can't we push back when people make remarks indicating that they find caring labor of all kinds, including birthing, intrinsically less important and intellectual than other kinds of labor? Why don't we ask why many people think of care as boring, soft, and submissive, not up to the standards or pace of our hard-edged time? And why do we so often prefer the opposite of care?

To reframe care within my own mind, I had to deal with my own suspicion that caring and mothering were somehow not intellectual or critically headed enough. This work did not conform to my idea of myself as a rigorous thinker, carrying the risk that I would fail myself by descending into a miasma of physical tasks. I even feared that having a baby attached to my person would somehow annihilate my own identity.

I have found just the opposite to be true. It began when I realized that I mothered more with my brain than with my body, and that this was just fine. My love story with my daughter began in earnest when I attached to her through our shared love of the dusty 1960s picture books that were antique even when they were read to me as a child, when my family lived in Great Britain. Reading them, my daughter and I floated together for hours in a sea of language and pictures.

I also reframed childrearing for myself by reading from the ethics of care, a strain of philosophy written about by scholars like Martha Nussbaum and Lisa Baraitser. A philosopher in this vein might argue, for instance, that a feminist ethics of care complicates the idea that certain individuals or groups are simply "autonomous," like a business elite, or simply "vulnerable," like women and caregivers and children, and instead shows what is inside these categories as well as their interdependence.

Baraitser, who studies precisely what I needed, maternal ethics, has written of the mother's unique perspective and her values of "interdependence, flexibility, relatedness, receptivity . . . preservative love"; she calls out these values as just the qualities we should all enshrine. Baraitser's analysis came on the heels of a similar concept, "maternal thinking," as defined by Sara Ruddick, a philosopher and mother. In 1989, Ruddick argued that motherhood involves distinct and highly desirable ways of thinking. Parenting is a discipline. It demands flexible thinking in relationship to children—whose brains Ruddick termed "open structures." What if being able to manage and lead minds so different from our grown-up minds is in fact *valuable* in not just a personal or familial way but also in a professional sense? In a global marketplace, after all, we are confronted by a wide range of minds and personalities. ("There was something so valuable about what happened when one became a mother," wrote the novelist Toni Morrison about this advantage. "Liberating because the demands that children make are not the demands of a normal 'other.' The children's demands on me were things that nobody else ever asked me to do. To be a good manager. To have a sense of humor.")

Wouldn't being able to manipulate and deal with other minds—a skill learned from our kids—be a clear asset at work? Anthropologists like Sarah Blaffer Hrdy also have argued for the motherhood advantage from a biological perspective, using primate moms as examples: Hrdy's animal mothers are nurturing but also ambitious, fighting for their and their young's survival, managing their own desires and ambivalence as well as their passion for their offspring.

Parenting, in either the biological or the ethics-of-care version, is not just a harried or degraded occupation. It requires a kind of mental elevation.

Perhaps a new concept of parenthood and care could help all parents create personal spaces of resistance and value. Mothers, fathers, care workers, and guardians hold deep knowledge about listening, reasoning, leadership, and scheduling. We've learned complex organization from planning pickups and drop-offs. We have a distinct understanding of other minds derived from comprehending our children's wildly versatile and often topsy-turvy thinking and behavior.

If we thought through care as a form of knowledge, we might recognize, as Daniela Nanau did in the first chapter of this book, that parenthood, rather than being the negative image of traditional work, actually prepares and even sharpens us for the workplace. Parenting elicits essential skills in us that prepare us for experiences beyond having children. If we were to embrace these "capabilities," we might see mothers as more than milk-stained wretches and fathers as more than squelched figures of fun.

Third, we can rethink the tenacity of traditional gender roles within our domestic partnerships. While male parents are doing more child care than they did in the past, according to the scholar Michael Kimmel, they aren't doing the scrubbing and toy clean-up as much as their female partners do. "Fathers are more likely to be taking their daughter to play ball while her mother cleans the house and makes lunch back home," as Kimmel puts it. In other words, women's huge contribution to parenting includes much of the household work, which is still invisible and out of balance. Making this hidden work more obvious and increasing male parents' contributions to household chores is a shift that we can go for on a familial level.

Finally, parents can start to *talk openly about social class*. This advice sounds both obvious and obscure, so what do I mean exactly? Just this: Let's treat our feelings about social class directly. Let's talk to our friends, our families, and our children about it. If our political and corporate cultures pit families against one another, we can at least start to resist that by being frank about our position and our feelings about our position.

Learning how to have these conversations is the work of organizations such as Class Action in the Boston neighborhood of Jamaica Plain. This educational nonprofit helps teachers and parents talk and think about social class. Its executive director, Anne Phillips, described its workshops and trainings as openly discussing "class cultures," or social class and income. Class Action's ultimate goal is what Phillips called "class sharing": open discussions in workshops about "who gets paid what and why, with people acknowledging their class backgrounds and the strengths and limitations that people have."

You might think this is just a fancy exercise in which those in the top classes "check their privilege," but Phillips assured me that though these programs operate mostly in private schools, they reach people from a range of class backgrounds.

Some schools have set about tackling inequality or income difference in the schoolyard. Pre-kindergarteners at New York City's Manhattan Country School, for instance, participate in a program that has them starting to talk about class (and ethnic and racial) lines and differences by visiting each other's houses and discussing their neighborhoods, what they have in their homes, and what kind of food they eat. At the Shady Hill School in Cambridge, Massachusetts, staff developed questions to guide similar work. "How do we help students understand differences, especially in a society that links value to wealth?" they asked. School families expressed concerns too—for instance, over the tendency of school benefits or potlucks to be held at the richest parents' houses.

If we could support more widespread school curricula about social class, we might discuss the full complexity of wealth within the parameters of our children's educational lives. Out of these lesson plans, we might talk more about what society values—and whether it rewards the right things. After all, a parent's profession may not be highly remunerative, but it might reflect training or skill. Attainment for some of us is not reflected in the dollars we accumulate (or at least not entirely) but in having our skills align with our work. For some families, status is about renown. For others, it's about avocation or craft.

These conversations are up to us as parents as well. I started to talk about class and money with my own six-year-old daughter as of this writing. She asked lots of questions, and I took a breath—I wanted to call Class Action to see what I should say next. I knew that these conversations were natural, and that it's a child's job to be curious. I won't lie: these conversations were difficult.

To start with, my daughter asked me why there were so many homeless people in our neighborhood—especially the homeless woman with the pet sparrow on her shoulder. My daughter then wanted us to give away all her snacks for the day—her cheddar bunnies and lemonade. Why do we have a home and they don't? she asked me. I explained that the homeless people may have had homes but had fallen out of whatever life they once led.

Then she wondered aloud why our apartment had no balcony and some of her friends' apartments did. Were they rich? Why didn't we have a balcony? My answer to that question was even more convoluted than my answer to her first question. I told her that some people make more money and others make less, but sometimes the people making less money choose to do so because they want to spend their time working with their hands or teaching.

Money and one of its embodiments, social class, are both riveting and mysterious to children. If we don't challenge today's stigma around speaking of familial class status, it will warp a new generation's experience of an even more important class—the places where they learn. And that's a distortion we simply can't afford.

We look to our political leaders, our courts, and our corporations to help get us out of the squeeze. As we wait, we can, of course, make further shifts within ourselves as well. Change shouldn't have to happen first internally and then later—someday—societally. This is a benighted order of things. But it is what we have now.

ACKNOWLEDGMENTS

Great literature should be "composed of the voices of life itself, from what I had heard in childhood, from what can be heard now in the street, at home, in a café, on a bus," at least according to my favorite writer as of late, the Russian author Svetlana Alexievich.

Similarly, it is to "the voices of life itself" that I owe my primary thanks for this book. I am most grateful to the American families who have shared their experiences with me and thus compose *Squeezed*.

After the subjects who revealed pieces of themselves to me, I must thank my inspired agent, Jill Grinberg, and her thoughtful colleague Denise St. Pierre. Jill has gone above and beyond her role as an agent in so many respects. Her insight is peerless, her critical intelligence a beacon.

Thirdly, I'd like to commend my fine editor, Denise Oswald. Denise contains a rare combination of good judgment and good values. I benefited immensely from her input on the book in many various stages and have also enjoyed our many conversations over the years. Appreciation also goes to her team at HarperCollins, including Emma Janaskie, and the publicity team of Ashley Garland, Meghan Deans, and Miriam Parker.

I'd like to recognize this book's peerless fact-checker, Queen Arsem-O'Malley, a great researcher. This book is so much better for her working on it.

Friends form a crucial lifeline for the subjects of *Squeezed*, and a social network is what keeps most journalists, including myself, afloat. This book would never exist without them. First among them is Maia Szalavitz, an alter ego and a collaborator. She read almost every draft and this book would never have existed without her sharp mind, generosity, and mysterious levels of niceness. Other confidants have been immensely helpful in the creation of this book. They are all talented writers themselves. They include gifted Ann Neumann and Abby Ellin, brave and witty women who were willing to go through drafts with felt-tipped pens. I also must recognize John Timpane: his editorial thoughts were astute and crucial.

I must also heap praise on the great Barbara Ehrenreich, who embraced me and brought me into her Economic Hardship Reporting Project four years ago and in the process transformed my life. Editing with her and stewarding EHRP, an incredible journalistic organization, has been a remarkable privilege. I truly love her, as a friend, colleague, journalist, and force of history.

I'd like to acknowledge as well my editors at various publications, primarily the gifted Jessica Reed, who edits my column at the *Guardian*. I'd also like to thank some of the editors at publications where I developed the early ideas that are now part of this book: Lizzy Ratner, Susan Lehman, Maria Streshinsky, Trish Hall, and Elizabeth Weil, among others. I am grateful for Capital & Main's Danny Feingold and Steven Mikulan's editorial assistance. I've also benefited from helpful colleagues at EHRP, who pitched in and worked all the harder when I went on mini–book leaves to complete this volume. Special credit goes to David Wallis for his skills and excellent common sense—which is never common—as well as to the utterly efficient Alexis Garcia and now Tanveer Ali.

I'd like to also offer encomium to the many, many close friends whom I have spoken to and leaned on while I wrote this book. My best friend since college, scholar Eleana Kim, read the manuscript and offered her oh-so-well-informed takes. Newer friends, scholars Elena Krumova and Devorah Baum, offered theoretical thoughts. Vicky de Grazia looked over the introduction with a welcome critical eye. Richard Kaye supplied reading suggestions. Jennifer Dworkin offered some philosophical direction. Other loyal and keen pals provided necessary feedback: Katherine Stewart, Helaine Olen, Rachel Urkowitz, Sarah Safer, Dale Maharidge, Deborah Siegel, Catherine Talese, Lauren Sandler, Becky Roiphe, Judith Matloff, Anne Kornhauser, Elizabeth Felicella, Astra Taylor, and Jared Hohlt. Additional props to Peter Mendelsund for his ever-brilliant cover advice and also David Opdyke for his input. In addition, I'd like to praise my favorite moms for their insta–focus groups regarding the book's cover, including Samantha Schonfeld, Melorra Sochet, and Kate Wagner-Goldstein. And thank you, Jon Segal, for starting this book off on its proverbial best foot.

In addition, in light of this book's subject matter and stance, I must thank our family's fantastic caregivers that made my admittedly squeezed work hours possible, first among them Sydney Viles but also Billy, Yoela, and Kylie and so many other remarkable young women and men.

Great gratitude also flows to my mother, Barbara Quart, whose long professional experience as an editor made this book better and whose own writing and feminist scholarship and early struggle to work, create, and mother simultaneously have served as an inspiration to me.

My dearest thanks of all go to my husband, Peter Maass, and my daughter, Cleo Quart Maass. Loving you both made me want to tell the stories of other families, and our story, a little, as well.

BIBLIOGRAPHY

Baraitser, Lisa. *Maternal Encounters: The Ethics of Interruption.* London: Routledge, 2009.

Beck, Richard. *We Believe the Children: A Moral Panic in the 1980s.* New York: PublicAffairs, 2015.

Bellow, Saul. *The Adventures of Augie March.* New York: Viking Press, 1953.

Berlant, Lauren. *Cruel Optimism.* Durham, NC: Duke University Press, 2011.

Bianchi, Suzanne, Nancy Folbre, and Douglas Wolf. "Unpaid Care Work." In *For Love and Money: Care Provision in the United States,* edited by Nancy Folbre, 40–65. New York: Russell Sage Foundation, 2013.

Bourdieu, Pierre. *Distinction: A Social Critique of the Judgement of Taste.* London: Routledge, 1986.

Brodkin, Karen. *Making Democracy Matter: Identity and Activism in Los Angeles.* New Brunswick, NJ: Rutgers University Press, 2007.

Brown, Tamara Mose. *Raising Brooklyn: Nannies, Childcare, and Caribbeans Creating Community*. New York: New York University Press, 2011.

Cusk, Rachel. *A Life's Work: On Becoming a Mother.* London: Picador, 2003.

De Beauvoir, Simone. *The Second Sex.* New York: Vintage, 2011.

Dreiser, Theodore. *The Financier.* New York: Harper & Brothers, 1912; reprint, New York: Penguin, 2008.

Ehrenreich, Barbara. *Fear of Falling: The Inner Life of the Middle Class.* New York: Pantheon Books, 1989.

Ehrenreich, Barbara, and Deirdre English. *For Her Own Good: Two Centuries of the Experts' Advice to Women.* New York: Random House, 2005.

Eribon, Didier. *Returning to Reims.* Translated by Michael Lucey. Los Angeles: Semiotexte, 2013.

Federici, Silvia. *Wages against Housework.* Bristol, U.K.: Falling Water Press, 1975.

Ford, Martin. *Rise of the Robots: Technology and the Threat of a Jobless Future.* New York: Basic Books, 2016.

Freeman, Joshua. *Behemoth: A History of the Factory and the Making of the Modern World.* New York: W. W. Norton & Company, 2018.

Goffman, Erving. *Stigma: Notes on the Management of Spoiled Identity.* New York: Touchstone, 1986.

Hill, Steven. *Raw Deal: How the "Uber Economy" and Runaway Capitalism Are Screwing American Workers.* New York: St. Martin's Press/Griffin, 2017.

Hirsch, Marianne. "Maternity and Rememory in Toni Morrison's *Beloved.*" In *Representations of Motherhood,* edited by Donna Bassin, Margaret Honey, and Meryle Mahrer Kaplan, 92–112. New Haven, CT: Yale University Press, 1996.

Hochschild, Arlie Russell. *The Outsourced Self: Intimate Life in Market Times.* New York: Metropolitan Books, 2012.

Hrdy, Sarah Blaffer. *Mother Nature: Maternal Instincts and How They Shape the Human Species.* New York: Ballantine Books, 2000.

Kessler-Harris, Alice. *Women Have Always Worked: A Historical Overview.* New York: Feminist Press, 1981.

Kimmel, Michael. *Angry White Men: American Masculinity at the End of an Era*. New York: Nation Books, 2013.

Kremer-Sadlik, Tamar, and Elinor Ochs, eds. *Fast-Forward Family: Home, Work, and Relationships in Middle-Class America.* Berkeley: University of California Press, 2013.

Kwak, Nancy H. *A World of Homeowners: American Power and the Politics of Housing Aid.* Chicago: University of Chicago Press, 2015.

Livingston, James. *No More Work: Why Full Employment Is a Bad Idea*. Chapel Hill: University of North Carolina Press, 2016.

Mittell, Jason. *Complex TV: The Poetics of Contemporary Television Storytelling.* New York: New York University Press, 2015.

Murphy, Finn. *The Long Haul: A Trucker's Tales of Life on the Road.* New York: W. W. Norton & Company, 2017.

Ngai, Sianne. *Ugly Feelings*. Cambridge, MA: Harvard University Press, 2007.

Noah, Timothy. *The Great Divergence: America's Growing Inequality Crisis and What We Can Do about It.* New York: Bloomsbury Press, 2013.

Nussbaum, Martha C. *Creating Capabilities: The Human Development Approach.* Cambridge, MA: Belknap Press of Harvard University Press, 2011.

Peck, Don. *Pinched: How the Great Recession Has Narrowed Our Futures and What We Can Do about It.* New York: Broadway Books, 2012.

Piketty, Thomas. *Capital in the Twenty-First Century.* Translated by Arthur Goldhammer. Cambridge, MA: Belknap Press of Harvard University Press, 2014.

Rich, Adrienne. *Of Woman Born: Motherhood as Experience and Institution.* New York: W. W. Norton & Company, 1995.

Rifkin, Jeremy. *The Age of Access: The New Culture of Hypercapitalism, Where All of Life Is a Paid-For Experience.* New York: TarcherPerigee, 2001.

Robinson, Fiona. *The Ethics of Care: A Feminist Approach to Human Security.* Philadelphia: Temple University Press, 2015.

Ruddick, Sara. *Maternal Thinking: Toward a Politics of Peace.* Boston: Beacon Press, 1995.

Sandage, Scott A. *Born Losers: A History of Failure in America.* Cambridge, MA: Harvard University Press, 2006.

Standing, Guy. *The Precariat: The New Dangerous Class.* London: Bloomsbury Academic, 2014.

Suárez-Orozco, Carola, and Marcelo M. Suárez-Orozco. *Children of Immigration.* Cambridge, MA: Harvard University Press, 2001.

Thomas, Gillian. *Because of Sex: One Law, Ten Cases, and Fifty Years That Changed American Women's Lives at Work.* New York: St. Martin's Press, 2016.

Tirado, Linda. *Hand to Mouth: Living in Bootstrap America.* New York: G. P. Putnam's Sons, 2014.

Walley, Christine J. *Exit Zero: Family and Class in Postindustrial Chicago.* Chicago: University of Chicago Press, 2013.

Zelizer, Viviana A. *The Purchase of Intimacy.* Princeton, NJ: Princeton University Press, 2007.

NOTES

INTRODUCTION

4 *the middle class is:* "America's Shrinking Middle Class: A Close Look at Changes within Metropolitan Areas," Pew Research Center, Washington, D.C., May 11, 2016, http://www.pewsocialtrends.org/2016/05/11/americas-shrinking-middle-class-a-close-look-at-changes-within-metropolitan-areas/.

5 *the cost of daily life:* To calculate the cost of a middle-class life, *Marketplace* considered the costs of the material goods associated with the middle class—such as vacations and homeownership—as well as the costs of basic needs like groceries, gasoline, and health care. Tommy Andres, "Does the Middle Class Life Cost More Than It Used To?" *Marketplace,* Minnesota Public Radio, June 9, 2016, https://www.marketplace.org/2016/06/09/economy/does-middle-class-life-cost-more-it-used.

5 *the price of a four-year degree at a public college:* "Fast Facts: Tuition Costs of Colleges and Universities," National Center for Education Statistics, Washington, D.C., 2016, https://nces.ed.gov/fastfacts/display.asp?id=76.

6 *after the* precariat: Guy Standing used the term in his 2011 book *The Precariat: The New Dangerous Class* (London: Bloomsbury Academic, 2014).

7 *the largest wealth inequality gap: Global Wealth Report 2015,* Credit Suisse Research Institute, October 2015, https://publications.credit-suisse.com/tasks/render/file/?fileID=F2425415-DCA7-80B8-EAD989AF9341D47E.

7 *Those born in the 1980s:* Raj Chetty et al., "The Fading American Dream: Trends in Absolute Income Mobility since 1940," *Science* 356, no. 6336: 398-406 (2017); David Leonhardt, "The American

Dream, Quantified at Last," *New York Times,* December 8, 2016, https://www.nytimes.com/2016/12/08/opinion/the-american-dream-quantified-at-last.html.

8 *Even those who were struggling:* Rakesh Kochhar and Rich Morin, "Despite Recovery, Fewer Americans Identify as Middle Class," Pew Research Center, Washington, D.C., January 27, 2014, http://www.pewresearch.org/fact-tank/2014/01/27/despite-recovery-fewer-americans-identify-as-middle-class/.

9 *according to a Pew study:* Ibid.

9 *A 2014 Russell Sage Foundation report:* Fabian T. Pfeffer, Sheldon Danziger, and Robert F. Schoeni, "Wealth Levels, Wealth Inequality, and the Great Recession," Russell Sage Foundation, New York, June 2014, http://web.stanford.edu/group/scspi/_media/working_papers/pfeffer-danziger-schoeni_wealth-levels.pdf.

CHAPTER 1

13 *In 2016, a report:* Cynthia Thomas Calvert, "Caregivers in the Workplace: Family Responsibilities Discrimination Litigation Update 2016," Center for WorkLife Law, University of California, Hastings College of the Law, San Francisco, May 2016, http://worklifelaw.org/publications/Caregivers-in-the-Workplace-FRD-update-2016.pdf, 4.

15 *pregnancy-related discrimination charges:* Vickie Elmer, "Workplace Pregnancy Discrimination Cases on the Rise," *Washington Post,* April 8, 2012, https://www.washingtonpost.com/business/capitalbusiness/workplace-pregnancy-discrimination-cases-on-the-rise/2012/04/06/gIQALWId4S_story.html?utm_term=.497c611809f3.

15 *if they take paternity leave:* Scott Coltrane, "The Risky Business of Paternity Leave," *The Atlantic,* December 29, 2013, https://www.theatlantic.com/business/archive/2013/12/the-risky-business-of-paternity-leave/282688/.

16 *they could be right:* "Parental Leave Survey," Deloitte, June 15, 2016.

16 *men who take on more of the caregiving:* Jennifer L. Berdahl and Sue H. Moon, "Workplace Mistreatment of Middle Class Workers Based on Sex, Parenthood, and Caregiving," *Journal of Social Issues* 69, no. 2 (2013): 341–66.

16 *employers are less likely to hire mothers:* Shelley J. Correll, Stephen Benard, and In Paik, "Getting a Job: Is There a Motherhood Penalty?" *American Journal of Sociology* 112, no. 5 (March 2007): 1297-339, http://gender.stanford.edu/sites/default/files/motherhoodpenalty.pdf.

16 *an average of $11,000 less:* Claire Cain Miller, "The Motherhood Penalty vs. the Fatherhood Bonus," *New York Times,* September 6, 2014, https://www.nytimes.com/2014/09/07/upshot/a-child-helps-your-career-if-youre-a-man.html.

17 *7 percent per child:* Michelle J. Budig and Paula England, "The Wage Penalty for Motherhood," *American Sociological Review* 66, no. 2 (2001): 204.

19 *through the American Civil Liberties Union:* "ACLU Files Discrimination Charges against Frontier Airlines on Behalf of Breast-Feeding Pilots," ACLU, New York, May 10, 2016, https://www.aclu.org/news/aclu-files-discrimination-charges-against-frontier-airlines-behalf-breast-feeding-pilots.

21 *After the births of her children, she says:* Rachel Cusk, *A Life's Work: On Becoming a Mother* (London: Picador, 2003), 5.

23 *the cost of delivery:* Tara Haelle, "Your Biggest C-Section Risk May Be Your Hospital," *Consumer Reports,* May 16, 2017.

23 *Only 14 percent of American workers:* "Paid Leave," National Partnership for Women & Families, Washington, D.C., http://www.nationalpartnership.org/issues/work-family/paid-leave.html.

24 *$2,669, and in 2015:* Mariko Oi, "How Much Do Women around the World Pay to Give Birth?" BBC News, February 13, 2015, http://www.bbc.com/news/business-31052665.

24 *In Quebec, Canada:* "Québec Parental Insurance Plan," Gouvernement du Québec, http://www4.gouv.qc.ca/EN/Portail/Citoyens/Evenements/DevenirParent/Pages/regm_quebc_assur_parnt.aspx.

24 *In Denmark:* Nancy Rasmussen, "Working in Denmark: Taking Parental Leave," The Local, May 6, 2015, https://www.thelocal.dk/20150506/working-in-denmark-maternity-and-parental-leave.

24 *Iceland's "daddy leave":* Dwyer Gunn, "How Should Parental Leave Be Structured? Ask Iceland," *Slate,* April 3, 2013, http://www.slate.com/blogs/xx_factor/2013/04/03/paternity_leave_in_iceland_helps_mom_succeed_at_work_and_dad_succeed_at.html.

25 *Pamela Druckerman's 2016* New York Times *essay:* Pamela Druckerman, "The Perpetual Panic of American Parenthood," *New York Times,* October 13, 2016, https://www.nytimes.com/2016/10/14/opinion/the-perpetual-panic-of-american-parenthood.html.

25 *only 13 percent of private-sector workers:* Drew Desilver, "Access to Paid Family Leave Varies Widely across Employers, Industries," Pew Research Center, Washington, D.C., March 23, 2017, http://www.pewresearch.org/fact-tank/2017/03/23/access-to-paid-family-leave-varies-widely-across-employers-industries/.

26 *90 percent pay, Canada:* "OECD Family Database," Organization for Economic Cooperation and Development, Paris, http://www.oecd.org/els/family/database.htm.

26 *they realize that their mothers can stay home:* To be sure, certain American states emulate the more bearable parenthood policies of Montreal or Copenhagen. In 2002, California started offering paid family leave benefits to eligible workers as part of the state's disability insurance program. Most workers have less than 1 percent of their wages withheld from each paycheck and put into family leave, as with Social Security. Workers pay into State Disability Insurance ("SDI" on pay stubs) and then get to withdraw their six weeks of benefits; benefit amounts range from $50 to $1,173 per week. Employees then siphon some of this money off when they need economic support for their new babies. They get only six weeks, though, and SDI covers only 55 percent of a worker's paycheck—not the oh-so-helpful leave offered in Sweden. Rhode Island, New York, and New Jersey have adopted a similar program, and there is pressure to make family leave a federal program.

28 *sociologist Erving Goffman:* Goffman uses the term "covering" to describe the behaviors or strategies an individual uses to manage stigmas. Goffman, *Stigma: Notes on the Management of Spoiled Identity* (New York: Touchstone, 1986).

28 *legal scholars Deborah L. Brake and Joanna L. Grossman:* Deborah L. Brake and Joanna L. Grossman, "Unprotected Sex: The Pregnancy Discrimination Act at 35," *Duke Journal of Gender Law and Policy* 21 (2013): 67.

29 *Joan Williams calls the "maternal wall":* Joan C. Williams, "The Glass Ceiling and the Maternal Wall in Academia," *New Directions for Higher Education* 130 (2005): 91.

CHAPTER 2

34 *Coverage for a child:* "Medicaid and CHIP Eligibility Levels," Medicaid, April 2016, https://www.medicaid.gov/medicaid/program-information/medicaid-and-chip-eligibility-levels/index.html.

35 *rose from 9,776 to 33,655:* These numbers were calculated by a senior researcher with the Urban Institute.

35 *This proportion was 8 percent:* Hope Yen, "Which Group Now Receives the Most Food Stamps in U.S.?" Associated Press, January 27, 2014, https://www.equalvoiceforfamilies.org/new-group-receives-the-most-food-stamps-in-u-s-who/.

36 *A survey of five hundred adjuncts:* Bettina Chang, "Survey: The

State of Adjunct Professors," *Pacific Standard,* March 19, 2015, https://psmag.com/economics/2015-survey-state-of-adjunct-professors.

37 *Employment for recent law school graduates:* Josh Mitchell, "Grad-School Loan Binge Fans Debt Worries," *Wall Street Journal,* August 18, 2015, https://www.wsj.com/articles/loan-binge-by-graduate-students-fans-debt-worries-1439951900.

39 *"exists when something you desire":* Lauren Berlant, *Cruel Optimism* (Durham, NC: Duke University Press, 2011), 1.

39 *"doing what you love" in America:* Miya Tokumitsu, *Do What You Love: And Other Lies about Success and Happiness* (New York: Regan Arts, 2015).

41 *only 24.1 percent:* Keith Hoeller, "The Wal-Mart-ization of Higher Education: How Young Professors Are Getting Screwed," *Salon,* February 16, 2014, http://www.salon.com/2014/02/16/the_wal_mart_ization_of_higher_education_how_young_professors_are_getting_screwed/.

44 *tested by a Princeton economist:* Dean Spears, "Economic Decision-Making in Poverty Depletes Behavioral Control" (working paper, Princeton University, 2010), 12.

45 *in her book* Hand to Mouth: Linda Tirado, *Hand to Mouth: Living in Bootstrap America* (New York: G. P. Putnam's Sons, 2015), 25.

48 *in a moment:* The 2016 election highlighted the anti-elite—and in particular, anti-intellectual—atmosphere that now prevails in certain quarters of America, what the scholar Devorah Baum sees as a new valorization of ignorance in our political life. While that strain could be called Trumpism, I'd argue that even more strikingly it represents the rise of the identity of "the uneducated." The Trump campaign being "uneducated" to be a quality that voters could be proud of. As Trump himself said after proving victorious in several caucuses: "I love the poorly educated." This championing of this group goes against decades of American thinking. Mainstream politicians have typically encouraged people to pursue higher education, the conventional wisdom of the previous century, valuing literacy and area expertise in both our leaders and citizens.

48 *At Yale University:* Alissa Quart, "Adventures in Neurohumanities," *The Nation,* May 8, 2013, https://www.thenation.com/article/adventures-neurohumanities/.

51 *In 2015:* "Earnings in the Past 12 Months (in 2015 Inflation-Adjusted Dollars)": 2015 American Community Survey One-Year Estimates," FactFinder, U.S. Census Bureau, Washington, D.C., https://factfinder.census.gov/faces/tableservices/jsf/pages/productview.xhtml?pid=ACS_15_1YR_S2001&prodType=table.

54 *more than 40 percent of the teachers:* Jordan Weissmann, "The Ever-Shrinking Role of Tenured College Professors (in 1 Chart)," *The Atlantic,* April 10, 2013, https://www.theatlantic.com/business/archive/2013/04/the-ever-shrinking-role-of-tenured-college-professors-in-1-chart/274849/.

54 *With a national median salary:* "A Portrait of Part-Time Faculty Members," Coalition on the Academic Workforce, June 2012, http://www.academicworkforce.org/CAW_portrait_2012.pdf.

54 *a pay target of $15,000:* Colleen Flaherty, "15K per Course?" Inside Higher Ed, February 9, 2015, https://www.insidehighered.com/news/2015/02/09/union-sets-aspirational-goal-adjunct-pay.

58 *PrecariCorps also received a request:* Alissa Quart, "The Professor Charity Case," *Pacific Standard,* March 19, 2015, https://psmag.com/social-justice/professor-charity-case-adjuncts-precaricorps.

58 *during National Adjunct Walkout Day:* "On National Adjunct Walkout Day, Professors Call Out Poverty-Level Wages and Poor Working Conditions," *Democracy Now!* February 25, 2015, https://www.democracynow.org/2015/2/25/on_national_adjunct_walkout_day_professors.

58 *according to U.S. Department of Education data:* Paul F. Campos, "The Real Reason College Tuition Costs So Much," *New York Times,* April 4, 2015, https://www.nytimes.com/2015/04/05/opinion/sunday/the-real-reason-college-tuition-costs-so-much.html.

59 *ten times the rate of growth:* John Hechinger, "The Troubling Dean-to-Professor Ratio," Bloomberg News, November 21, 2012, https://www.bloomberg.com/news/articles/2012-11-21/the-troubling-dean-to-professor-ratio.

59 *That bill:* Suzanne Hudson, "SB15-094 Defeated in Committee," American Association of University Professors Colorado Conference, January 29, 2015, https://aaupcolorado.org/page/2/.

60 *the "alienated prison":* Max Weber, *The Protestant Ethic and the Spirit of Capitalism,* trans. Talcott Parsons (New York: Dover, 2003).

CHAPTER 3

65 *30 percent of Americans:* G. William Domhoff, "The Rise and Fall of Labor Unions in the U.S.," Who Rules America? University of California, Santa Cruz, February 2013, http://www2.ucsc.edu/whorulesamerica/power/history_of_labor_unions.html; Megan

Dunn and James Walker, "Union Membership in the United States," U.S. Bureau of Labor Statistics, Washington, D.C., September 2016, https://www.bls.gov/spotlight/2016/union-membership-in-the-united-states/pdf/union-membership-in-the-united-states.pdf.

65 *nearly 40 percent of Americans:* Harriet B. Presser, "The Economy That Never Sleeps," *Contexts* 3, no. 2 (2004): 42.

65 *The average American adult:* Lydia Saad, "The '40-Hour' Workweek Is Actually Longer—by Seven Hours," Gallup News, August 29, 2014, http://news.gallup.com/poll/175286/hour-workweek-actually-longer-seven-hours.aspx.

66 *the number of part-timers working:* Ben Casselman, "Yes, Some Companies Are Cutting Hours in Response to 'Obamacare,'" FiveThirtyEight, January 13, 2015, https://fivethirtyeight.com/features/yes-some-companies-are-cutting-hours-in-response-to-obamacare/.

68 *According to U.S. Bureau of Labor Statistics:* "Employment Projections," U.S. Bureau of Labor Statistics, Washington, D.C., https://data.bls.gov/projections/occupationProj.

70 *one such company, Kid Care Concierge:* Kid Care Concierge, http://kidcareconcierge.com/.

70 *As Rachel Cusk writes* Rachel Cusk, *A Life's Work: On Becoming a Mother* (New York: St. Martin's Press/Picador, 2001), 145.

71 *17 percent of American workers:* "At Least 17 Percent of Workers Have Unstable Schedules," Economic Policy Institute, Washington, D.C., April 9, 2015, http://www.epi.org/press/at-least-17-percent-of-workers-have-unstable-schedules/.

71 *Despite a federal bill:* "Recently Introduced and Enacted State and Local Fair Scheduling Legislation," National Women's Law Center, Washington, D.C., May 2015, https://www.nwlc.org/sites/default/files/pdfs/recently_introduced_and_enacted_state_and_local_fair_scheduling_legislation_apr_2015.pdf.

71 *Eighteen percent work sixty:* Saad, "The '40-Hour' Workweek."

72 *Extreme day care:* The scholars Paula England and Nancy Folbre have written on care workers developing emotional attachments to their clients—children and other vulnerable populations—and feeling as if they could hurt their clients by demanding higher wages. Meanwhile, the owners and managers who run hospitals, day cares, and schools never meet the cared-for populations and are thus able to unemotionally cut costs and restructure hours. And indeed, it might well "hurt" parents who are themselves often stretched thin to advocate for better conditions and demand higher wages.

72 *the United States is close to the bottom:* "PF3.1: Public Spending on Childcare and Early Education," Organization for Economic Cooperation and Development, Social Policy Division, Directorate of Employment, Labour, and Social Affairs, November 22, 2011, https://www.oecd.org/els/soc/PF3_1_Public_spending_on_childcare_and_early_education.pdf, 2.

73 *the cost of center-based day care:* Elise Gould and Tanyell Cooke, Issue Brief 404: "High Quality Child Care Out of Reach for Working Families," Economic Policy Institute, Washington, D.C., October 6, 2015, https://www.childcaresolutionscny.org/sites/default/files/child-care-is-out-of-reach.pdf, 2.

76 *when most relationships:* Jeremy Rifkin, *The Age of Access: The New Culture of Hypercapitalism Where All of Life Is a Paid-For Experience* (New York: TarcherPerigee, 2001), 112.

77 *The ideal, she writes:* Viviana A. Zelizer, *The Purchase of Intimacy* (Princeton, NJ: Princeton University Press, 2007), 22.

80 *In Finland, all children:* Claudio Sanchez, "What the U.S. Can Learn from Finland, Where School Starts at Age 7," *Weekend Edition,* NPR, March 8, 2014, http://www.npr.org/2014/03/08/287255411/what-the-u-s-can-learn-from-finland-where-school-starts-at-age-7.

80 *in Canada, the province of Quebec:* "Quebec Daycare Fees to Climb to $20 per Day for Highest-Earning Families," CBC News, November 20, 2014, http://www.cbc.ca/news/canada/montreal/quebec-daycare-fees-to-climb-to-20-per-day-for-highest-earning-families-1.2841994.

81 *Colorado's licensed day-care spots:* "2014 Signature Report: An Analysis of Colorado's Licensed Child Care System," Qualistar, September 2014, https://issuu.com/qualistarcolorado/docs/2014_qualistar_colroado_signature_r, 3; Megan Verlee, "Report: Colorado Faces Statewide Daycare Shortage," Colorado Public Radio, September 25, 2014.

81 *the number of in-home child-care providers:* Kirsti Marohn, "Shortage of Affordable Child Care Hampers Families," *Saint Cloud Times,* April 2, 2016, http://www.sctimes.com/story/news/local/2016/04/02/shortage-affordable-child-care-hampers-families/82262546/.

83 *"The program is so popular":* Dana Goldstein, "Bill de Blasio's Pre-K Crusade," *The Atlantic,* September 7, 2016, https://www.theatlantic.com/education/archive/2016/09/bill-de-blasios-prek-crusade/498830/?utm_source=feed.

84 *According to a Hechinger Institute report:* W. Steven Barnett et al. "The State of Preschool 2016," National Institute for Early Education Research, Rutgers Graduate School of Education, New

Brunswick, NJ, 2017, http://nieer.org/wp-content/uploads/2017/05/YB2016_StateofPreschool2.pdf.

86 *28 percent of children:* "Single Parenthood in the United States—A Snapshot (2014 Edition)," Women's Legal Defense and Education Fund, New York, https://www.legalmomentum.org/sites/default/files/reports/SingleParentSnapshot2014.pdf, 1.

CHAPTER 4

90 *only if it also improves people's social rank:* Christopher J. Boyce, Gordon D. A. Brown, and Simon C. Moore, "Money and Happiness: Rank of Income, Not Income, Affects Life Satisfaction," *Psychological Science* 21, no. 4 (2010): 471.

91 *Glenn Firebaugh and Matthew Schroeder:* Glenn Firebaugh and Matthew B. Schroeder, "Does Your Neighbor's Income Affect Your Happiness?" *American Journal of Sociology* 115, no. 3 (2009): 805.

92 *"People's understanding of prosperity":* danah boyd, "Failing to See, Fueling Hatred," *Wired*, March 3, 2017, https://www.wired.com/2017/03/dont-hate-silicon-valley-techies-who-complain-about-money/.

93 *associate professor in behavioral science:* Michael Daly, Christopher Boyce, and Alex Wood, "A Social Rank Explanation of How Money Influences Health," *Health Psychology* 34, no. 3 (2015): 222.

94 *Edith Wharton once described:* Edith Wharton, *The Age of Innocence* (New York: D. Appleton & Company, 1920; reprint, CreateSpace, 2015), 32.

95 *Neal Gabler, who in 2016:* Neal Gabler, "The Secret Shame of Middle-Class Americans," *The Atlantic*, May 2016, https://www.theatlantic.com/magazine/archive/2016/05/my-secret-shame/476415/.

95 *Helaine Olen described:* Helaine Olen, "All the Sad, Broke, Literary Men," *Slate*, April 21, 2016, http://www.slate.com/articles/business/the_bills/2016/04/neal_gabler_s_atlantic_essay_is_part_of_an_old_aggravating_genre_the_sad.html.

96 *known as the Whitehall Studies:* M. G. Marmot et al. "Health Inequalities among British Civil Servants: The Whitehall II Study," *The Lancet* 337, no. 8754 (June 8, 1991): 1387–93, http://www.thelancet.com/journals/lancet/article/PII0140-6736(91)93068-K/abstract.

97 *an enviable array of illiquid assets:* Justin Weidner, Greg Kaplan, and Giovanni Violante, "The Wealthy Hand-to-Mouth," *Brookings Papers on Economic Activity* (Spring 2014): 77–153, https://www.brookings.edu/wp-content/uploads/2016/07/2014a_Kaplan.pdf.

97 *spending money to hire tutors:* Adair Morse and Marianne Bertrand, "Trickle-Down Consumption," *Review of Economics and Statistics* 98, no. 5 (2016): 863.

99 *The EPI also found:* Elise Gould, Tanyell Cooke, and Will Kimball, "What Families Need to Get By: EPI's 2015 Family Budget Calculator," Economic Policy Institute, Washington, D.C., August 26, 2015, http://www.epi.org/publication/what-families-need-to-get-by-epis-2015-family-budget-calculator/.

101 *University of California, Berkeley, has famously written:* Emmanuel Saez, "U.S. Top One Percent of Income Earners Hit New High in 2015 amid Strong Economic Growth," Washington Center for Equitable Growth, July 1, 2016, http://equitablegrowth.org/research-analysis/u-s-top-one-percent-of-income-earners-hit-new-high-in-2015-amid-strong-economic-growth/.

101 *The Equality of Opportunity Project:* Raj Chetty et al., "The Fading American Dream: Trends in Absolute Income Mobility since 1940," *Science* 356, no. 6336 (2017): 398–406.

102 *Leichter, an attorney and writer:* Debra Cassens Weiss, "At Least Half of the Lawyers in These Nine States and Jurisdictions Aren't Working as Lawyers," ABA Journal, June 1, 2017, http://www.abajournal.com/news/article/at_least_half_of_the_lawyers_in_these_nine_states_and_jurisdictions_arent_w.

103 *as the* New York Times *reports:* Steven J. Harper, "Too Many Law Students, Too Few Legal Jobs," *New York Times,* August 25, 2015, https://www.nytimes.com/2015/08/25/opinion/too-many-law-students-too-few-legal-jobs.html?mcubz=1.

103 *American student debt overall:* Zack Friedman, "Student Loan Debt in 2017: A $1.3 Trillion Crisis," *Forbes*, February 21, 2017, https://www.forbes.com/sites/zackfriedman/2017/02/21/student-loan-debt-statistics-2017/#3e097bfc5dab.

103 *"from less advantaged classes":* Didier Eribon, *Returning to Reims,* trans. Michael Lucey (Los Angeles: Semiotexte, 2013).

105 *Paul Campos wrote of Florida Coastal:* Paul Campos, "The Law-School Scam," *The Atlantic,* September 2014, https://www.theatlantic.com/magazine/archive/2014/09/the-law-school-scam/375069/.

107 *the very cheapest law schools:* Ilana Kowarski, "U.S. News Data: Law School Costs, Salary Prospects," *U.S. News & World Report,* March 15, 2017, https://www.yahoo.com/news/u-news-data-law-school-costs-salary-prospects-130000280.html.

107 *The study was called, pungently:* Deborah Jones Merritt and Kyle McEntee, "The Leaky Pipeline for Women Entering the Legal Profession," November 2016, https://www.lstradio.com/women/documents/MerrittAndMcEnteeResearchSummary_Nov-2016.pdf.

109 *the journalist Tim Noah puts it:* Tim Noah, "The United States of Inequality," *Slate*, September 14, 2010.

CHAPTER 5

112 *the "global care chain":* Arlie Hochschild has used this term in her work since 2000.

113 *places like the Mississippi Delta:* "Poverty-Stricken Past and Present in the Mississippi Delta," *PBS NewsHour*, July 22, 2016, http://www.pbs.org/newshour/bb/poverty-stricken-past-present-mississippi-delta/.

113 *Harrington would have said:* Michael Harrington, *The Other America: Poverty in the United States* (New York: Touchstone, 1962), 14.

116 *8.2 million children under five:* "Who's Minding the Kids? Child Care Arrangements: 2011—Detailed Tables," U.S. Census Bureau, Washington, D.C., April 13, 2013, https://www.census.gov/data/tables/2008/demo/2011-tables.html.

118 *According to a 2012 study:* "Home Economics: The Invisible and Unregulated World of Domestic Work," National Domestic Workers Alliance, 2016, https://www.2016.domesticworkers.org/homeeconomics/.

118 *the cost of child care:* The labor protections that were part of the New Deal's National Labor Relations Act of 1935 allowed private-sector employees to unionize. Domestic workers, almost certainly due to racism, were excluded. Southern lawmakers also insisted that job protections for farmworkers and domestic workers, many of whom were black, should be excluded from President Franklin D. Roosevelt's Fair Labor Standards Act of 1938. Informal workers were also denied social benefits, like minimum wage, and unemployment compensation. Eighty years later, on a state level, things are now changing somewhat. In 2010, for instance, New York enacted the Domestic Workers' Bill of Rights, which establishes an eight-hour workday and other protections.

118 *"When you say you have kids":* Ai-jen Poo has been leading her campaign to reframe care work, as she said in one interview, "from poverty-wage work that is invisible and undervalued into professional jobs with opportunities for career advancement." As Poo notes, care jobs are growing five times as fast as other common jobs. Caregiving is the spine that supports American economic productivity, yet informal labor doesn't advance workers the way manufacturing jobs did during the last century.

123 *over 1 million family-based immigration petitions:* "Number of Form I-130, Petition for Alien Relative, by Category, Case Status, and USCIS Field Office or Service Center Location," U.S. Citizenship and Immigration Services, January 1–March 31, 2017, https://www.uscis.gov/sites/default/files/USCIS/Resources/Reports%20and%20Studies/Immigration%20Forms%20Data/Family-Based/I130_performancedata_fy2017_qtr2.pdf.

128 *the scholar Paula England:* England, "Emerging Theories of Care Work," 381.

129 *A 2015 McKinsey report:* Jonathan Woetzel et al., "The Power of Parity: How Advancing Women's Equality Can Add $12 Trillion to Global Growth," McKinsey Global Institute, September 2015, https://www.mckinsey.com/global-themes/employment-and-growth/how-advancing-womens-equality-can-add-12-trillion-to-global-growth.

129 *American parents, it is estimated:* Alexandra Sifferlin, "Women Are Still Doing Most of the Housework," *Time,* June 18, 2014, http://time.com/2895235/men-housework-women/.

129 *Mothers do fifteen more:* Kim Parker and Eileen Patten, "The Sandwich Generation: Rising Financial Burdens for Middle-Aged Americans," Pew Research Center, Washington, D.C., January 30, 2013, http://www.pewsocialtrends.org/2013/01/30/the-sandwich-generation/.

129 *The BLS American Time Use Survey:* "Table A-1. Time Spent in Detailed Primary Activities and Percent of the Civilian Population Engaging in Each Activity, Averages per Day by Sex, 2016 Annual Averages," U.S. Bureau of Labor Statistics, Washington, D.C., 2016, https://www.bls.gov/tus/a1_2016.pdf.

129 *American artist Mierle Laderman Ukeles wrote:* Mierle Laderman Ukeles, "Manifesto for Maintenance Art, 1969," Arnolfini, https://www.arnolfini.org.uk/blog/manifesto-for-maintenance-art-1969.

137 *A recent report out of the University of California, Los Angeles:* John Kucsera, with Gary Orfield, *New York State's Extreme School Segregation: Inequality, Inaction and a Damaged Future,* Civil Rights Project/Proyecto Derechos Civiles, March 2014, https://files.eric.ed.gov/fulltext/ED558739.pdf.

137 *Studies have found:* Sam Roberts, "Gap between Manhattan's Rich and Poor Is Greatest in U.S., Census Finds," *New York Times,* September 17, 2014, https://www.nytimes.com/2014/09/18/nyregion/gap-between-manhattans-rich-and-poor-is-greatest-in-us-census-finds.html.

139 *the income of the top 1 percent:* "Inequality," *The State of Working America,* Economic Policy Institute, Washington, D.C., http://www.stateofworkingamerica.org/fact-sheets/inequality-facts/.

139 *as calculated by Larry Summers:* Lawrence Summers, "It Can Be Morning Again for the World's Middle Class," *Financial Times,* January 18, 2015, https://www.ft.com/content/826202e2-9d85-11e4-8946-00144feabdc0.

143 *As of 2015:* "DP02: Selected Social Characteristics in the United States: 2015 American Community Survey 1-Year Estimates: New York City and Boroughs," U.S. Census Bureau, Washington, D.C., 2015, http://www1.nyc.gov/assets/planning/download/pdf/data-maps/nyc-population/acs/soc_2015acs1yr_nyc.pdf.

143 *"By any measure, immigration":* Carola Suárez-Orozco and Marcelo M. Suárez-Orozco, *Children of Immigration* (Cambridge, MA: Harvard University Press, 2001).

CHAPTER 6

148 *Uber Oregon unrolled an app:* "Support Portland Metro Area Schools This Summer," *Uber Blog,* June 20, 2016, https://www.uber.com/blog/portland/support-portland-metro-area-public-schools-this-summer/.

150 *"precarious manhood" theory:* Joseph A. Vandello and Jennifer K. Bosson, "Hard Won and Easily Lost: A Review and Synthesis of Theory and Research on Precarious Manhood," *Psychology of Men and Masculinity* 14, no. 2 (2013): 101.

150 *male studies scholar Michael Kimmel*: Michael Kimmel, *Angry White Men: American Masculinity at the End of an Era* (New York: Nation Books, 2013).

152 *teachers in the San Francisco Unified School Distric:t* Heather Knight and Joaquin Palomino, "Teachers Priced Out: SF Educators Struggle to Stay amid Costly Housing, Stagnant Salaries," *San Francisco Chronicle,* May 13, 2016, http://projects.sfchronicle.com/2016/teacher-pay/.

155 *Emmanuel Levinas writes:* Emmanuel Levinas, *On Escape,* trans. Bettina Bergo (Stanford, CA: Stanford University Press, 1935), 63.

155 *"Uber has been extremely clever":*

157 Richard M. Ingersoll, "High Turnover Plagues Schools," September 2002, *USA Today*, August 14, 2002, retrieved from http://repository.upenn.edu/gse_pubs/130

159 *every third worker a freelancer:* Susan Adams, "More Than a Third of U.S. Workers Are Freelancers Now, but Is That Good for Them?" *Forbes,* September 5, 2014, https://www.forbes.com/sites/susanadams/2014/09/05/more-than-a-third-of-u-s-workers-are-freelancers-now-but-is-that-good-for-them/#54754dd921c3.

161 *the judge rejected the settlement:* Davey Alba, "Judge Rejects

Uber's $100 Million Settlement with Drivers," *Wired*, August 18, 2016, https://www.wired.com/2016/08/uber-settlement-rejected/.

CHAPTER 7

167 *a French film I had seen:* The novelist Emmanuel Carrere described this particular plot as a horror movie for the downsized: "He spent his days wandering around, avoiding his neighborhood. He doesn't speak to anyone, every face frightens him because it could belong to a former colleague." Emmanuel Carrere, *The Adversary,* (Picador).

168 *more than 60 percent of Americans:* "2016 Student Loan Data Update," Center for Microeconomic Data, Federal Reserve Bank of New York, New York, https://www.newyorkfed.org/microeconomics/databank.html.

168 *as one real estate developer did:* Same Levin "Millionaire Tells Millennials: If You Want a House, Stop Buying Avocado Toast," *Guardian,* May 15, 2017, https://www.theguardian.com/lifeandstyle/2017/may/15/australian-millionaire-millennials-avocado-toast-house.

168 *a paper by the Federal Reserve Bank of New York:* "Household Debt and Credit: 2017 Q2 Report," Center for Microeconomic Data, Federal Reserve Bank of New York, New York, https://www.newyorkfed.org/microeconomics/hhdc/background.html.

169 *according to a study on long-term unemployment:* Patricia Cohen, "Over 50, Female and Jobless Even as Others Return to Work," *New York Times,* January 1, 2016, https://www.nytimes.com/2016/01/02/business/economy/over-50-female-and-jobless-even-as-others-return-to-work.html.

169 *the Equal Employment Opportunity Commission:* "Charge Statistics (Charges Filed with EEOC) FY 1997 through FY 2016," U.S. Equal Employment Opportunity Commission, https://www.eeoc.gov/eeoc/statistics/enforcement/charges.cfm.

169 *long-term unemployment rates for people:* Karen Kosanovich and Eleni Theodossiou Sherman, "Trends in Long-Term Unemployment," U.S. Bureau of Labor Statistics, Washington, D.C., March 2015, https://www.bls.gov/spotlight/2015/long-term-unemployment/pdf/long-term-unemployment.pdf.

177 *The school shut down suddenly:* Shahien Nasiripour, "ITT Technical Institute Shuts Down, Leaving a Hefty Bill," Bloomberg News, September 6, 2016, https://www.bloomberg.com/news/articles/2016-09-06/itt-technical-institutes-shuts-down-leaving-a-hefty-bill.

178 *over conning students into believing:* Camila Domonoske, "Judge Approves $25 Million Settlement of Trump University Lawsuit," NPR, March 31, 2017, http://www.npr.org/sections/thetwo-way/2017/03/31/522199535/judge-approves-25-million-settlement-of-trump-university-lawsuit.

178 *sham real estate "flipping" seminars:* Patrick Danner, "San Antonio House-Flipper Montelongo Sued by 164 Ex-Students," *San Antonio Express-News,* March 2, 2016, http://www.expressnews.com/real-estate/article/San-Antonio-house-flipper-Montelongo-sued-by-164-6866991.php.

179 *"modern, a low hum":* Scott Sandage, *Born Losers: A History of Failure in America* (Cambridge, MA: Harvard University Press, 2006), 256.

180 *"I once thought that there were no second acts":* F. Scott Fitzgerald, "My Lost City," in *My Lost City: Personal Essays, 1920-1940,* ed. James L. W. West III (Cambridge: Cambridge University Press, 2005), 114.

180 *"I am an American, Chicago born":* Saul Bellow, *The Adventures of Augie March* (New York: Viking Press, 1953), 1.

180 *"A real man—a financier—is never a tool":* Theodore Dreiser, *The Financier* (New York: Harper & Brothers, 1912; reprint, New York: Penguin, 2008), 20.

180 *the American Everyman Failure:* Or in the words of the character Biff Loman: "Well, I spent six or seven years after high school trying to work myself up. Shipping clerk, salesman, business of one kind or another. And it's a measly manner of existence. To get on that subway on the hot mornings in summer . . . and always to have to get ahead of the next fella." Arthur Miller, *Death of a Salesman* (Harmondsworth, U.K.: Penguin Books, 1996), 22.

181 *The "ruthless self-centeredness":* Barbara Ehrenreich and Deirdre English, *For Her Own Good: Two Centuries of the Experts' Advice to Women* (New York: Random House, 2005), 331.

182 *Columbia Law School labor specialist Mark Barenberg writes:* Mark Barenberg, "Widening the Scope of Worker Organizing: Legal Reforms to Facilitate Multi-Employer Organizing, Bargaining, and Striking," Roosevelt Institute, New York, October 7, 2015, http://rooseveltinstitute.org/wp-content/uploads/2015/10/Widening-the-Scope-of-Worker-Organizing.pdf.

184 *In 2015, the Consumer Financial Protection Bureau:* Lance Williams, "How Corinthian Colleges, a For-Profit Behemoth, Suddenly Imploded," Reveal, September 20, 2016, https://www.revealnews.org/article/how-corinthian-colleges-a-for-profit-behemoth-suddenly-imploded/.

184 *Jobs programs must focus:* There are also more conventional ways to alleviate the suffering of the middle-aged, out-of-work parent, like publicly funded and targeted job training instead of the jumbled salmagundi so many encounter when they try to find their second and third acts on their own. This is a different *kind* of job training: less individualistic, more practical, and focused on filling needs in the workforce as opposed to more personal and random ideas about a trade-centered education. Our country could fund more programs that support job training on the federal or state level and encourage more unions and nonprofits to also help create more apprenticeship programs. At the moment, the United States spends far less than other developed countries on job training systems and retraining. In his second term, President Obama tried to implement a German-style apprenticeship program to fill the continued demand for machinists and robotics specialists. A variation of such a system is currently being implemented under ApprenticeshipUSA, which has received $90 million in funding from the federal government to support state apprenticeships, especially in new industries.

Around San Antonio, Texas, for instance, ApprenticeshipUSA and Workforce Solutions Alamo supply individuals with apprenticeships as bricklayers or automotive technicians, both jobs that are in demand, and work to place people in these positions. (Perhaps as part of its monopoly on all things, Amazon is getting in on the apprenticeship program: the U.S. Department of Labor is training veterans for tech jobs at, yes, Amazon.)

CHAPTER 8

198 *immense reductions of newsroom staffs:* Ken Doctor, "Newsonomics: The Halving of America's Daily Newsrooms," Newsonomics, July 28, 2015, http://newsonomics.com/newsonomics-the-halving-of-americas-daily-newsrooms/.

200 *it was easier to cover rent:* Claude S. Fischer, "Reversal of Fortune," *Boston Review,* June 20, 2016, https://bostonreview.net/us/claude-fischer-reversal-fortune-urbanization-gentrification.

200 *urban scholar David "DJ" Madden:* David Madden and Peter Marcuse, *In Defense of Housing: The Politics of Crisis* (New York: Verso, 2016).

201 *Rent stabilization and control:* Rent control started in New York City in 1969 when rents really began to jack up in postwar buildings; today one million apartments are covered by these guidelines, which protect tenants from big rent increases. Some think that rent stabilization helps create a fairer housing market, protecting it from gentrification. Others argue that the price cap on these dwellings reduces supply, thus raising prices around the stabilized or controlled units. Some local governments and the federal government have tried to respond to the demand for affordable housing through legislation, such as offering tax incentives to building owners who house lower-income tenants.

201 *cohousing arose:* "Cohousing in the United States: An Innovative Model of Sustainable Neighborhoods," Cohousing Association of the United States, Boulder, CO, March 6, 2017, http://www.cohousing.org/sites/default/files/attachments/StateofCohousingintheU.S.%203-6-17.pdf.

204 *called motherhood a state:* Adrienne Rich, *Of Woman Born: Motherhood as Experience and Institution* (W. W. Norton & Company, 1995).

CHAPTER 9

209 *yesteryear's* Upstairs, Downstairs: The servants in *Upstairs, Downstairs* are certainly more appealing and compelling than the inhabitants of the pristine "upstairs" precinct; indeed, the first version of the program carried traces of the New Left class consciousness in Britain of that period. Not so coincidentally, the feminist novelist Fay Weldon wrote the first episode of *Upstairs, Downstairs,* emphasizing the maids' point of view.

209 *the most common American leisure activity:* "American Time Survey: Leisure and Sports Activities: Leisure Time on an Average Day," U.S. Bureau of Labor Statistics, Washington, D.C., last modified December 20, 2016, https://www.bls.gov/TUS/CHARTS/LEISURE.HTM.

211 *"The vast wealth depicted":* For example, in a meta-touch, the $150 million Netflix series *The Crown,* which is devoted to Queen Elizabeth II and the royal family, features an episode about the first televised coronation: in sharing its excessive pageantry with the citizenry via TV, the monarchy distracted them from their postwar rations and hardships.

212 *A decade ago, in a* New York Times *article:* Felicia R. Lee, "Being a Housewife Where Neither House nor Husband Is Needed," *New York Times,* March 5, 2008, http://www.nytimes.com/2008/03/05/arts/television/05real.html.

212 *about $2 billion in free media exposure:* Nicholas Confessore and Karen Yourish, "$2 Billion Worth of Free Media for Donald Trump," *New York Times*, March 15, 2016, https://www.nytimes.com/2016/03/16/upshot/measuring-donald-trumps-mammoth-advantage-in-free-media.html.

213 *Will Wilkinson wrote:* Will Wilkinson, "The Majesty of Trump," *New York Times,* November 2016, https://www.nytimes.com/interactive/projects/cp/opinion/election-night-2016/the-majesty-of-trump.

214 *the writer Michael Lerner:* Michael Lerner, "Stop Shaming Trump Supporters," *New York Times,* November 9, 2016, https://www.nytimes.com/interactive/projects/cp/opinion/election-night-2016/stop-shaming-trump-supporters.

215 *Manovich studied millions of social media images:* Lev Manovich and Alise Tifentale, "Our Main Findings," Selfiecity, http://www.selfiecity.net/#findings.

216 *"By posting selfies":* Mark R. Leary, "Scholarly reflections on the 'Selfie,'" OUPblog, November 19, 2013, https://blog.oup.com/2013/11/scholarly-reflections-on-the-selfie-woty-2013/. Leary was asked, along with several other scholars, to write a short reflection; the article is not attributed to him.

216 *"people can convey":* "Prosumption," originally coined by Alvin Toffler, combines production and consumption in the same action. We offer ourselves on social media, to our friends, glamorized and sanitized. Somewhere therein is a hope that these images will raise our general social capital and will cover over, say, our downsized journalism or legal careers.

216 *the wealthy of television past:* If we go even further back, before television ever flickered into view, the films of one of the worst economic times in our history, the Depression, were famous for their opulence. Poor Americans lined up in droves to see screen sirens in flouncy dresses and cheery musical numbers. When they did address inequality then, it was through films like *My Man Godfrey*—whose central character, a hobo, is whisked off the streets to a party of that era's 1 percent, at the behest of a capricious wealthy woman—or *Citizen Kane*—whose well-heeled Charles Foster Kane runs a newspaper conglomerate.

217 *"adult-minded serials":* Thomas Doherty, "Storied TV: Cable Is the New Novel," *Chronicle of Higher Education,* September 17, 2012, http://www.chronicle.com/article/Cable-Is-the-New-Novel/134420.

218 *household incomes nationwide:* "America's Shrinking Middle Class: A Close Look at Changes within Metropolitan Areas," Pew Research Center, Washington, D.C., May 11, 2016, http://www.pewsocialtrends.org/2016/05/11/americas-shrinking-middle-class-a-close-look-at-changes-within-metropolitan-areas/.

220 *TV was originally considered:* Michael Z. Newman and Elana

Levine, *Legitimating Television: Media Convergence and Cultural Status* (New York: Routledge, 2012).

CHAPTER 10

227 *The American Trucking Associations says:* American Trucker Associations, "Reports, Trends & Statistics," http://www.trucking.org/NewsandInformationReportsIndustryData.aspx

227 *projected a total loss of 7.1 million jobs:* World Economic Forum, "The Future of Jobs: Employment, Skills and Workforce Strategy for the Fourth Industrial Revolution," January 2016, p.13.

228 *spending nearly $1 million to research:* Joelle Renstrom, "Robot Nurses Will Make Shortages Obsolete," Daily Beast, September 24, 2016.

228 *a 2013 McKinsey Global Institute report:* James Manyika et al., "Disruptive Technologies: Advances That Will Transform Life, Business, and the Global Economy," McKinsey Global Institute, May 2013, https://www.mckinsey.com/business-functions/digital-mckinsey/our-insights/disruptive-technologies.

228 *2016 survey by Evans Data Corporation:* Evans Data Corporation, "Software Developers Worry They Will Be Replaced By AI," March 2016.

228 *a 2015 study by Ball State University:* Michael J. Hicks and Srikant Devaraj, "The Myth and the Reality of Manufacturing in America," Ball State University, Muncie, IN, June 2015, http://projects.cberdata.org/reports/MfgReality.pdf.

228 *A recent report by a market research company:* Brian Hopkins et al., "The Top Emerging Technologies to Watch: 2017 to 2021," Forrester, September 12, 2016, https://www.forrester.com/report/The+Top+Emerging+Technologies+To+Watch+2017+To+2021/-/E-RES133144.

230 *The first commercial delivery:* Marco della Cava, "Self-Driving Truck Makes First Trip—A 120-Mile Beer Run," *USA Today,* October 26, 2016, https://www.usatoday.com/story/tech/news/2016/10/25/120-mile-beer-run-made-self-driving-truck/92695580/.

231 *today's drivers are themselves*: "My Problem With Uber All Along." http://www.rushkoff.com/rebooting-work/ Douglas Rushkoff's Website, Also Medium, October 17, 2015 in a review "Getting Over Uber" Susan Crawford.

234 *Mardi Thompson, a nurse:* Jenny Gold, "The Orderly Zipping around the Hospital May Be a Robot," *Marketplace*, Minnesota

Public Radio, February 23, 2016, https://www.marketplace.org/2016/02/25/health-care/orderly-zipping-around-hospital-may-be-robot.

235 *The Bureau of Labor Statistics projects:* "Occupational Outlook Handbook: Registered Nurses," U.S. Bureau of Labor Statistics, Washington, D.C., last modified October 24, 2017, https://www.bls.gov/ooh/healthcare/registered-nurses.htm.

236 *"robot"-written stories:* Benjamin Mullin, "Robot-Writing Increased AP's Earnings Stories by Tenfold," Poynter, January 29, 2015, https://www.poynter.org/news/robot-writing-increased-aps-earnings-stories-tenfold.

237 *theorist Zeynep Tufekci agrees:* Zeynep Tufekci, "Failing the Third Machine Age: When Robots Come for Grandma," *Medium*, July 22, 2014, https://medium.com/message/failing-the-third-machine-age-1883e647ba74.

238 *the three hundred-pound Robear:* James Griffiths, "Singapore Turns to Robots to Get Seniors Moving," CNN, February 29, 2016, http://www.cnn.com/2015/10/20/asia/singapore-aging-robot-coaches-seniors/index.html.

238 *the chipper 2014* New York Times *op-ed:* Louise Aronson, "The Future of Robot Caregivers," *New York Times*, July 19, 2014, https://www.nytimes.com/2014/07/20/opinion/sunday/the-future-of-robot-caregivers.html.

238 *theorist Sianne Ngai puts it:* Sianne Ngai, "Theory of the Gimmick," *Critical Inquiry* 43, no. 2 (2017): 467.

239 *MIT's Julie Shah and her coauthors:* Adam Conner-Simons, "Robot Helps Nurses Schedule Tasks on Labor Floor," MIT News, July 13, 2016. http://news.mit.edu/2016/robot-helps-nurses-schedule-tasks-on-labor-floor-0713.

240 *their mechanical rivalry:* The tech intellectual Jaron Lanier calls the winners of the rise of the robots the "siren servers." He dubs this tech ruling class—those who produce and finance these machines—"narcissists; blind to where value comes from, including the web of global interdependence that is at the core of their own value." (Those who reap the most from this efficiency are, by and large, rich technologists.) Jaron Lanier, *Who Owns the Future?* (New York: Simon & Schuster, 2013.)

241 *up from $691 billion in 2012:* Deborah Bach, "Study Reveals Surprising Truths about Caregivers," *UWNews*, June 16, 2015, https://www.washington.edu/news/2015/06/16/study-reveals-surprising-truths-about-caregivers/.

242 *The* Ottawa Citizen *kvelled:* Madeline Ashby, "Ashby: Let's Talk about Canadian Values (Values Like a Universal Basic Income),"

Ottawa Citizen, November 15, 2016, http://ottawacitizen.com/opinion/columnists/ashby-lets-talk-about-canadian-values-values-like-a-universal-basic-income.

242 *feminist theorist Kathi Weeks:* Kathi Weeks, *The Problem with Work: Feminism, Marxism, Antiwork Politics, and Postwork Imaginaries* (Durham, NC: Duke University Press, 2011).

242 *"no longer socially necessary":* James Livingston, *No More Work: Why Full Employment Is a Bad Idea* (Chapel Hill: University of North Carolina Press, 2016).

244 *the journalist Judith Shulevitz wrote:* Judith Shulevitz, "It's Payback Time for Women," *New York Times,* January 10, 2016.

247 *"impossible professions":* Barbara Almond, "The Fourth Impossible Profession," *Psychology Today,* November 5, 2010, https://www.psychologytoday.com/blog/maternal-ambivalence/201011/the-fourth-impossible-profession.

248 *eke out our income:* As the writer Sue Halpern put it in the *New York Review of Books:* "All economies have winners and losers. It does not take a sophisticated algorithm to figure out that the winners in the decades ahead are going to be those who own the robots, for they will have vanquished labor with their capital." Halpern, "Our Driverless Future," *New York Review of Books,* November 24, 2016, http://www.nybooks.com/articles/2016/11/24/driverless-intelligent-cars-road-ahead/.

CONCLUSION

252 *Stanford University professor Jerry Kaplan:* Jerry Kaplan, *Humans Need Not Apply: A Guide to Wealth and Work in the Age of Artificial Intelligence* (New Haven, CT: Yale University Press, 2015).

252 *A 2015 McKinsey study:* Michael Chui, James Manyika, and Mehdi Miremadi, "Four Fundamentals of Workplace Automation," *McKinsey Quarterly,* November 2015, https://www.mckinsey.com/business-functions/digital-mckinsey/our-insights/four-fundamentals-of-workplace-automation.

261 *"interdependence, flexibility, relatedness":* Lisa Baraitser, *Maternal Encounters: The Ethics of Interruption* (London: Routledge, 2009), 26.

261 *"There was something so valuable":* Toni Morrison, *Conversations with Toni Morrison,* ed. Danille K. Taylor-Guthrie (Jackson: University Press of Mississippi, 1994), 270.

262 *"Fathers are more likely":* Michael Kimmel, *Angry White Men: American Masculinity at the End of an Era* (New York: Nation Books, 2013).

INDEX

READING GROUP GUIDE

1. Which people or characters in the book do you identify with and why? Who didn't you identify with? Did anyone's story move you in particular?

2. What is the American dream to you? Is it home and car ownership or is it not having debt or raising a nuclear family? Having an advanced degree? Working a white-collar job? Or do you have another definition? Do you think it's harder to be a middle-class parent today than it was when you were growing up? How does this make you feel?

3. The concept of self-blame threads through *Squeezed*. Why do you think people blame themselves for social ills? Should they instead question their government or their company, health insurer, or university? Or are they right to assume so much personal responsibility?

4. Quart questions the "do what you love" philosophy that led a number of her book's subjects to financial peril. Is it better to be alienated from one's work yet be able to pay one's bills? Or is it better to love what you do yet struggle to earn your keep? Which of these trajectories reflect your own experience? Is it worse for millennials, in your opinion?

5. Middle-class life is roughly 30 percent more expensive now than in the mid-90s. Why is this so, according to *Squeezed*? Were you surprised to learn through the book that college professors, trained lawyers, and IT workers sometimes struggle to get by?

6. Do you feel comfortable discussing your economic situation with your friends and colleagues? With your children, if you have children? What do you think about Quart's notion that we should talk openly about our social class and financial instability? Would this work for you personally, or do you have another way to explain poverty, affluence, and wealth and monetary anxiety to kids?

7. Quart argues that America doesn't care about care. What does this phrase mean to you personally? Do you think caregiving—by parents, guardians, day-care workers, and babysitters—and even teachers and nurses—is undervalued in our country? If so, why?

8. As Quart shows, many European countries have paid maternity leave and subsidized childcare and the U.S. has far too little. How has this near-absence of paid leave and government-supported childcare in America impacted you and your family's life?

9. Quart writes about parents that attempt to hack the system through coparenting collectives, retraining programs, and bartering and trading, among other smaller fixes; and basic income guarantees among the larger ones. Which of these appeal to you personally? And what are your family's hacks that help you survive economically?

10. Quart coins the term "the motherhood advantage" to describe the parents discovering their leadership and workplace skills through parenting rather than in spite of it. How has your work life gotten better and worse since you became a parent? What insights has having a child given you that you bring to work each day?

11. Why did Quart focus mostly on middle-class parents rather than working-poor parents? Why are so many of the parents in this book women?

12. Why are Americans so attached to the classic idea of the middle class, which Quart shows only partly exists in its postwar form?

13. Quart finds new terms to describe the sometimes quite hard lives of people in this book, including "the forever clock" and "the middle precariat." What do these phrases mean to you?

14. Is this a moment in America when we could, in fact, devolve into a deeply troubled society? How did the election of President Donald Trump change the overall situation described in this book?

15. Some think of precarious employment or gig jobs as liberation. What do Quart and the people in this book think of this definition of independence? What might better unions and more of them—and more powerful unions or campaigns—potentially do for Americans who are not making enough or lack job security?

FURTHER READING

Refund, Karen Bender, 2015. If *Squeezed* were a book of short stories, it would read like this.

Raising Brooklyn: Nannies, Childcare, and Caribbeans Creating Community, Tamara Mose Brown, 2011. This scholarly volume very much informed how I wrote and thought about the caregivers in *Squeezed* and "the global care chain."

Lines the Quarry, Robin Clarke, 2013. This is the poetry of inequality. One poem in Clarke's book, for instance, is composed of a list that reflects her work as a union organizer: "917 assemblers & fabricators/150 athletes & sports competitors." The list poem ends with hundreds of registered nurses and thousands of retail workers. It's not a labor pool assembled out of poetic fancy but rather 2006 workers' compensation injury data. Powerful.

Evicted: Poverty and Profit in the American City, Matthew Desmond, 2017. Incredibly dense fieldwork makes this a crucial volume and a gripping read. Desmond also has a strong theory he is forwarding here: eviction as a key moment of crisis that then sets off a cascade of further social horrors.

Fear of Falling: The Inner Life of the Middle Class, Barbara Ehrenreich, 1989. This earlier iteration of the story of Americans afraid of downward mobility is a sterling prequel to what you've just read in these pages. Today, the fear of falling engulfs more and more people, a nervousness that helps define our period.

Wages Against Housework, Silvia Federici, 1975. This book emerged out of the feminist Wages for Housework movement of the 1970s. It is having a surprising renaissance, thanks in part to the sheer amount of unpaid work so many women (as well as men) now do, from laboring in the home to posting on Facebook. In the recent past, comical hashtags give a light modern spin to Federici's book, like the Twitter hashtag #GiveYourMoneyToWomen.

The Outsourced Self: What Happens When We Pay Others to Live Our Lives for Us, Arlie Russell Hochschild, 2013. This book describes how what was once part of private life, including child rearing, is now outsourced, and then sold back to desperate consumers as high-end expertise. At its most absurd, "the outsourced self" includes the nameologists, who you can pay to name your child.

White Collar: The American Middle Classes, C. Wright Mills, 1951. A classic study of the American middle class. It delineated the white-collar worker, and their social alienation. The existential unrest within *White Collar*'s middle-class workers may seem quaint today, caused by what Mills sees as a salesmanship mentality, among other things. After all, now, so many of us are salesmen—from our gig work to our social media presence—that self-marketing is merely the beginning of our problems.

This Is Not a House, Edgar Martins, 2011. This book of photographs taken after the home mortgage crisis depicts a broad anxiety about our dwellings in a time when so many in America have lost theirs. The abandoned homes could even be said to conjure haunted houses—they reveal a realm where nothing really belongs to itself.

Dead Pledges: Debt, Crisis, and Twenty-First-Century Culture, Annie McClanahan, 2016. This critical theory book seeks to make sense of personal debt from an aesthetic vantage. McClanahan looks at art and literature about the affliction of owing in modern life. She argues that debt-dependent works of art and literature oscillate in their mood "between comedy and pathos."

Unequal Childhoods: Class, Race, and Family Life, Annette Lareau, 2011. As clarifying as when it was first published, this book thickly describes the radical differences between the way children grow up in working poor families and wealthier ones.

Citizen: An American Lyric, Claudia Rankine, 2014. Rankine has found lyric and experimental ways to expose American contemporary society. She is definitive of what I think of as "civic poetry," where poets render economic, racial, and other kinds of political experience in vibrant, subjective, and irruptive ways.

Heartland: A Memoir of Working Hard and Being Broke in the Richest Country on Earth, Sarah Smarsh, 2018. At Economic Hardship Reporting Project, I edited a number of Sarah's pieces. Her approach definitely inflected *Squeezed*. *Heartland* toggles between a distinctive and lush memoir of growing up working poor and cogent analysis of the shame and denial Americans have around class.

NW, Zadie Smith, 2012. The novel's central character, Leah Hanwell, is certainly a squeezed person, swimming in student loans she has yet to pay off. According to the periodical *Dissent*, *NW* is the first novel "as far as I know, to deal directly with the consequences of student debt."

Working: People Talk About What They Do All Day and How They Feel About What They Do, Studs Terkel, 1997. A comprehensive account of the inner lives of working people, from steel laborers to dentists.

The House of Mirth, Edith Wharton, 1905. American novels where the central character is afflicted by social class immobility and/or debt compel me. These novels also animate my nonfiction writing on these subjects and, generally, I get a great deal of inspiration from the American realist and naturalist traditions. That's where Lily Bart comes in. In this great novel, Bart's initial "credit" is her beauty, yet she descends nonetheless into a deep debt that becomes part of her tragedy.